OBSERVING VARIABLE STARS

OBSERVING VARIABLE STARS

A guide for the beginner

David H. Levy

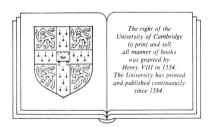

The right of the
University of Cambridge
to print and sell
all manner of books
was granted by
Henry VIII in 1534.
The University has printed
and published continuously
since 1584.

Cambridge University Press

Cambridge
New York   New Rochelle   Melbourne   Sydney

Published by the Press Syndicate of the University of Cambridge
The Pitt Building, Trumpington Street, Cambridge CB2 1RP
32 East 57th Street, New York NY 10022, USA
10 Stamford Road, Oakleigh, Melbourne 3166, Australia

First published 1989

Printed in Great Britain at the University Press, Cambridge

British Library cataloguing in publication data available

Library of Congress cataloguing in publication data available

ISBN 0 521 32113 1

HC

To my Mother, Father, and Grandparents, who have shared many starlit nights with me.

Contents

x   *Contents*

Foreword

Variable star astronomy is one field in which an amateur astronomer can still make significant contributions to science. Regardless of the optical tool used, whether it is the naked eye, binoculars, a small or a large telescope, any lover of the stars can play an important role in our understanding of variable stars.

David Levy is truly a lover of stars. He is an avid observer and a discoverer of four comets, the second one found only forty minutes after he finished the final draft of this book! I have known David for over a decade as a member of the AAVSO and as a friend. His enthusiasm and his exuberance for astronomy has always impressed me. When David talks about variable stars, it is as if he is talking about his friends; they are not just stellar objects.

David is keenly aware of the difficulties that a new variable star observer faces. He knows well that, in the beginning, locating variables and estimating their brightnesses takes lots of patience and perseverance. He also knows the joy one feels in making variable star observations. Therefore, he makes every effort to find ways to get his reader interested in variable stars, and to make that first brightness estimate. He helps his reader to find out more about the sky, the wonderful seasonal progression of its appearance, about astronomy as it applies to variable stars, and finally more about the different types of variable stars and the individual members of each type.

David's book is an expression of love of the sky and of variable stars. I hope you will take the journey with David into this wonderful field, read about it, explore it, go out and observe, and feel the joy and the satisfaction of knowing that through your observations you are helping solve some of the mysteries of these stars.

Janet A. Mattei
Director, American Association of Variable Star Observers

Opening thoughts

What do you see when you look through a telescope? Is it the mountains and valleys of a lunar highland, or perhaps a thinly-veiled Jovian storm? Or do you prefer the ghostly light of the distant galaxies, island universes adrift in a sea of space and time? Perhaps you see the fluctuations of stars in our galaxy, stars of all ages whose nightly appearance changes according to some cosmic drumbeat whose rhythm we try to unravel.

A variable star is simply a star that changes in brightness. Observing variable stars is both useful to science, and fun. It is a field that needs the observations that dedicated amateurs with small telescopes have the time and enthusiasm to make. It will reciprocate as you contribute to it, for the more you observe the more you will learn about your subjects of observation.

The purpose of this book is to inspire you to observe variable stars. Through its pages, I want to share my enthusiasm for these distant suns that change in brightness. Accordingly the book's approach is to emphasize the observing, and to keep the scientific explanations simple and in the background.

Why do variables attract us? The answer lies in the stars themselves. These innocuous objects attract our interest because of their behavior, not their appearance. Although they do not look fascinating at first glance, consistent watching will draw out their surprises.

Variable stars reward the patient observer. A cycle of observing almost any Mira star will show why these stars are so interesting. As a long period variable fades, it requires more and more telescope power to keep in touch. Week after week the star keeps on getting dimmer, giving you a feeling that you are about to witness its disappearance forever. Then one week you detect that the star's fading is slowing. You increase your frequency of observation to once every four days. Then comes a night when the star is actually brighter than before; it has passed its minimum and is now brightening. With each passing week its presence becomes more commanding as its reddish color stands out more and more among the neighbors in its field. In its approach to maximum, a Mira shines with pride.

When variable star observing became popular in the early years of this century, the prospect of amateur observers adding something to our understanding of nature was the main attraction. In his hugely popular *Field Book of the Skies*, William Tyler Olcott invited his readers to become part of "the great work of astrophysical research" through a program that could be accomplished from their own back yards.

The stellar wind of research changes in direction. Where the careful monitoring of hundreds of long period variable stars once was viewed as the major interest area for variables, today we follow other types as well. In 1920, few astronomers even knew about the stars that have periodic outbursts, the dwarf novae, which today are an important research field. A program for beginners today includes a mixture of Mira-type stars, dwarf novae, eclipsing binary stars and other stars whose light fluctuations are worth noting. Astronomy is a dynamic, evolving science, and the types of variables we add to our programs reflect this changing scene.

To say that 10% or 50% of variable star astronomers consider amateur work worthwhile is meaningless. For all those astronomers who use visual data, our work is invaluable. Some of the rest may think that visual data are not accurate enough. This is true, and data obtained from an electronic light-measuring device known as a photometer are more accurate and thus more valuable than data we get using our eyes. However, many of the stars we observe simply do not get watched by professionals with photometers or their more modern counterparts, the light-gathering chips we call charge-coupled devices or CCDs. Thus, our observations are the only available record of the behavior patterns of many stars. In that sense, visual data carefully taken are infinitely better than no data at all.

In learning the periods and the long term behavior of many variable stars, the visual observations we make are sufficient. In three-quarters of a century, the American Association of Variable Star Observers (AAVSO) has gathered near-continuous records of the behavior of many stars. When astronomers wish to learn the case history of a star, when they want to plan their observing programs, they rely on these data.

Quite aside from being objects of interest, the cyclic patterns are fun to watch. We observe variables for science, but also for sport. After a night of variable star observing we feel good. Our list of numbers shows that we have looked at the sky, and taken its pulse.

Getting started can be an experience. The first time I spent a night outside trying to estimate the brightness of a variable star, I resolved never to look at one again. The star I chose for that ill-fated night was Chi Cygni, a famous long period variable, which at maximum can often be seen clearly. But on this night, Chi Cygni was nowhere near maximum, and at 13th magnitude, it was completely lost in a sea of very faint Milky Way stars. I wasn't sure that Chi Cygni would be very bright that night, but I certainly didn't expect it to be as faint as that.

Yet today, this uncertainty about how bright a variable is going to be has become the main reason I keep observing them. They always offer surprises. Occasionally, an unexpectedly faint minimum will remind me of that long night with Chi Cygni.

I begin observing each night in a mood of suspense. What has changed since the last time I was outside? The first thing I do is to take a quick glance over the entire sky to see if any new novae, or exploding stars, have appeared. Tonight I shall check the sky down to the 3rd magnitude, just to see.

But wait! What's there? On the long arm of Cygnus — an extra star? Yes, it's my old friend Chi Cygni again, now bright and visible without even binoculars. Chi is an example of a long period variable.

In a year, I can watch a number of such stars pass through complete cycles of variation, occasionally stretching through as much as eight magnitudes on each side. Although the maximum and minimum of long period variables can be predicted with some confidence up to a year in advance, it is up to amateur observers to spot humps and small standstills in the light curves of some of these stars.

These are variable stars, and we are their watchers. Serious observing is like playing or composing music. To get the most out of these activities takes heart and soul. The mere thought of doing it gives you a pleasant feeling and a twinkle in your eye, but when you first put eye to eyepiece, or finger to keyboard, or pen to paper, your whole being is filled with a special satisfaction. With variables, this joy has the added dimension of being a part of what is happening away from home.

Acknowledgments

It is an honor to acknowledge the help of the following people:

Judy Stowell, for the enormous amount of time she spent typing and reading the manuscript, and for her meticulous work with the finder charts;

Steve Edberg, for his thorough review of the final draft;

Janet A. Mattei, Director of the American Association of Variable Star Observers, for her careful suggestions and encouragement;

James V. Scotti, for his expert and original creation of the finder charts and for some of the detail charts;

Isabel K. Williamson, for developing the Big Dipper brightness project (page 8) and some of the material in chapter 6;

David J. Eicher and Robert Burnham of *Deep Sky* magazine, for their help with earlier versions of some chapters;

Astronomy, *Deep Sky* and *Star and Sky* magazines, in which earlier versions of some chapters appeared;

Brian Skiff, for assistance in checking factual information;

Michael E. Bakich, for his original software and all-sky charts;

Simon Mitton and Cambridge University Press, and subeditor Sue Glover, whose assistance was invaluable;

John W. Griesé III, Peter Collins, and Mark Heifner, respected observers whose assistance was important;

Norman Sperling, for his editing of the manuscript, and Tom Glinos, whose typing and advice were very useful in its early stage;

Peter Jedicke, who wrote "The Stars go Nova" and "Betelgeuse, Betelgeuse", and who assisted with the Betelgeuse chapter;

Damien Lemay, Terence Dickinson, Jim Gall, Colleen Ryder, Gord Graham, Clifford Holmes, Leo Enright, Jeanine Cockrell, Richard Berry, and Perry Remaklus.

To all these, my grateful thanks for help, encouragement and suggestions, all appreciated more than I ever can say.

1

Getting to know the sky

1 Beginning with the Big Dipper

The best way to get a good start on observing is to go outside and discover the stars for yourself. Before you learn about observing variable stars, get your bearings and learn your way around the sky. Becoming familiar with it is an important first step toward useful observation.

We do need a place and time to start, so let's try your back yard, under an evening sky of late spring or early summer. High in the west will shine the seven bright stars of the Big Dipper, possibly the best known asterism, or group of stars, in the entire sky. Since Roman times they have been part of Ursa Major (UMa), the Greater Bear. The Dipper's handle represents the tail of the Bear, while the feet and nose are shown by fainter stars to the south and west of the bowl (Fig. 1.1).

Fig. 1.1. North circumpolar stars.

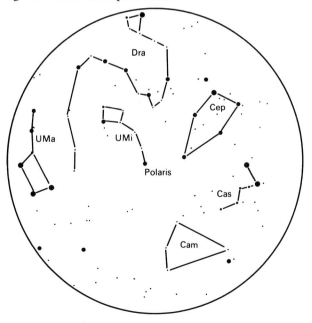

1.1   The Dipper as a road sign

Now, look carefully. Can you really see a bear up there? I've never come close to seeing one, even though I have tried. One night, while observing in the mountains of eastern Pennsylvania, I was assured by a friend that I would indeed have an encounter with a bear. Trying to look through a hazy sky, I heard a heavy sound of bush cracking, and I wondered what kind of bear was in store for me that night. As it turned out, I saw neither.

At any time of night and in any season of the year, the two stars at the end of the Dipper's bowl point towards Polaris, the North Star. All the stars in our sky, the Sun included, circle the celestial poles, and for at least the next 100 years, Polaris will stay within a degree of the true North Celestial Pole. Polaris

Fig. 1.2. The sky, January to March.

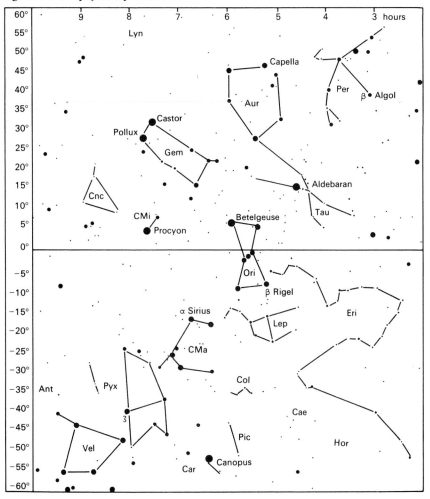

is the brightest star of another constellation, Ursa Minor or the Little Bear. This Little Dipper is bent back, resembling a spoon bent by a child. Possibly the tantrum was caused by the child's fruitless attempts to find the Little Dipper, for the figure is hard to see. Except for Polaris itself and the two stars at the end of the bowl, the Little Dipper's stars are faint.

We can use the Big Dipper to locate several nearby constellations. In northern hemisphere spring, try the handle, which curves, like an arc of a circle, to the southeast. Start by joining the stars of the handle with an imaginary curved line, and "arc" to Arcturus, a bright yellow-orange star in the constellation of Bootes (Fig. 1.3). Although this constellation is known mythologically as the Herdsman, it is easier to identify as a kite. Once you have seen Arcturus, why don't you keep your curved line moving in the same direction, for further along you will find Spica, a

Fig. 1.3. The sky, April to June.

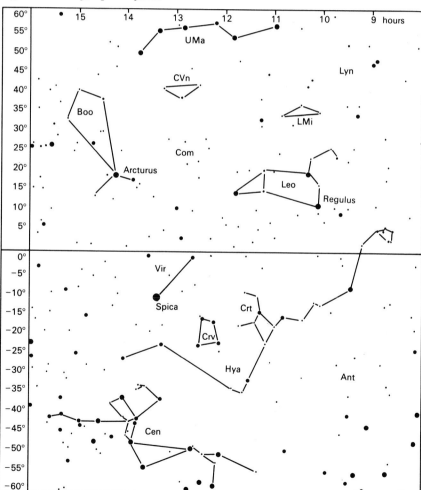

bright bluish star that is at the head of Virgo the Virgin. Next, draw a line from Gamma in the bowl through Eta, the end star of the handle, and continue across most of the summer sky towards Antares in Scorpius (Fig. 1.4). If you join the the two inner stars of the bowl (Gamma and Delta Ursae Majoris) with a line that continues north, you will eventually reach the summer triangle of Vega, Deneb, and Altair (Fig. 1.4.).

In fall, join the two stars at the end of the handle, Eta and Zeta, and continue above the bowl to Capella (Fig. 1.2). In late winter, when the Dipper is high in the sky, a line from Eta through Gamma will lead across the sky in the direction of Procyon. This brings us back to spring, when we also can use the Dipper in a different way. Fill the Dipper's bowl with water, and then poke some holes through the bottom. As the water

Fig. 1.4. The sky, July to September.

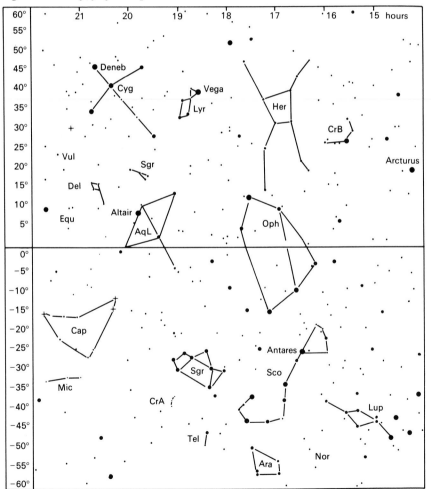

gushes out, note who is sitting below, comfortably showering himself: it's Leo the Lion! The bright star at the foot of the backward question mark that forms Leo's head is called Regulus (Fig. 1.3).

1.2   Your own constellations

Thousands of years ago the Egyptians, Greeks, and Romans placed their heroes in the sky. These figures are still with us because they offer an easy and familiar way of finding our bearings. Before you learn the standard constellations, why not make some up for yourself? This way, you can become familiar with the sky on your own terms, requiring no

Fig. 1.5. The sky, October to December.

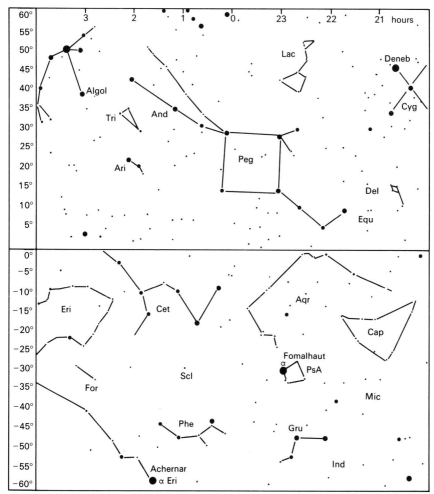

textbooks or charts; at this stage you are better off without them. And what is most surprising of all, when you begin to learn the classical groups you may be surprised at how closely your personal circus resembles that of tradition.

After a night or two of this you should be ready to begin your voyage to the constellations. You will now need a good star chart or atlas. Learning the sky by going from one constellation to another is a process that requires patience and will occupy the starlight nights of at least four seasons. The stars of spring with the Lion and the Crab give way as the year progresses to the Swan and the Harp of summer, and with the cooling air of autumn comes Pegasus, the mighty flying horse and Andromeda, the princess in chains. Finally, the two hunting dogs of Orion romp through the snowy sky of winter. Then, on late winter nights you can look towards the east and see the spring stars on display once more as another orbit of your life comes full circle. As you proceed, you may find that learning the constellations is like making new friends who will be there to greet you on schedule at their appointed times.

1.3   Southern Cross

Just as the Big Dipper points the way to so many northern hemisphere constellations, Crux, the Southern Cross, guides you to constellations around the south pole. When it is high in the southern

Fig. 1.6. South circumpolar stars.

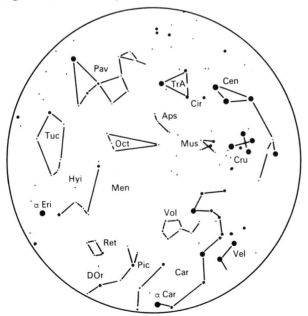

hemisphere autumn sky, follow the short arm towards Centaurus, and the long arm toward the pole and its faint constellations of Musca and Octans. Canopus and the delights of Carina also grace the fall sky.

The southern sky is a busy place, with more bright stars than northern hemisphere observers are used to. The Milky Way is magnificent, particularly in the southern part of Carina, where stars and clouds of gas combine to paint a pretty celestial picture. Our two nearest galaxies, the Large and Small Magellanic Clouds, are also visible only from southern latitudes, the Large Cloud in fall, the Small one in spring.

If you have ever seen the constellation of Sagittarius on a hot July evening under a dark sky, you probably noticed the Milky Way in its background, bright as a cloud. Imagine how glorious Sagittarius looks from the southern hemisphere, where it can be seen directly overhead at that time of year! The reason the Milky Way is so bright in that part of the sky is that Sagittarius marks the central hub of our Milky Way galaxy. If you live in the southern hemisphere, you are lucky to have such beauty in your back yard. If you live in the northern hemisphere, travel south some day. Some stars are there waiting to meet you.

2   Magnitude, color, and distance

What is the first thing you notice about the stars? Quite likely, it is their differing brightnesses. Although this appears obvious, it is the single most important concept with which you should become familiar before you can be a variable star observer.

2.1   Magnitude

Why do stars differ in brightness? Is it because they are at different distances from us, so that the farther stars appear fainter, like the glow from lamps at the far end of the street? Or are the stars themselves of different brightnesses? As you have likely guessed, both are correct. Sirius, a blue star off the southeast corner of Orion the Hunter, is normally the brightest star in the night sky, not so much because it is actually large and bright, but because it is close. At a mere 8 light years away, Sirius is one of our nearest neighbors. However, there exists a star only a little farther away, Wolf 359, whose intrinsic brightness is so low that it cannot be seen with the unaided eye. Meanwhile, the brightest star in Cygnus the Swan, Deneb, is well over 1000 light years away from us. S Doradus, intrinsically one of the brightest stars of all, appears to us as a faint star shining dimly because it is so far away, actually in a neighboring galaxy.

To reconcile these two factors, astronomers have created two independent systems of describing brightness or "magnitude." A star's "apparent magnitude" is a number that indicates its brightness as it shines in our sky. A star of the 2nd magnitude, like Polaris, is about 2.5 times fainter than a 1st magnitude object, like Aldebaran. Similarly, a 3rd magnitude star would be 2.5 times fainter than Polaris, and so on. Thus a 3rd magnitude star would be more than 6 times fainter than Aldebaran, a magnitude 4 star would be more than 15 times fainter, a magnitude 5 star 39 times fainter, and a magnitude 6 star would be 100 times fainter. This scale can, of course, apply for stars brighter than first magnitude too. Vega, a star about 2.5 times brighter than Aldebaran, is about magnitude 0, while brilliant Sirius shines at around magnitude -1.4.

While this system is useful in assigning star brightnesses as they appear to us on Earth, it does not tell us how intrinsically bright a star really is. For this purpose, we can use another system of "absolute magnitudes." Imagine a sky in which every object is at precisely the same distance from us, and let that distance be 33 light years. At this distance, the Sun would appear almost as bright as a star of the 5th magnitude. S Doradus would be extremely bright at magnitude -20, mighty Sirius would be extremely diminished, to magnitude 4. Such a sky would be odd indeed, for eyes used to the stars as we see them. It would not be filled as ours is, with stars that have unfairly taken places too close to the front of the room, and appear brighter than they deserve!

A project

Which of the seven stars in the Dipper is the brightest? Is it one of the three in the handle, or is it in the bowl? Which one is the faintest? (That should be easy.) And finally, what is the order in between? Do this exercise carefully: have your friends do so as well and keep them trying until you are fairly certain of the order. But watch out! You may be surprised to find that during a hazy or moonlit night the order appears to change.

2.2   Color

Stars show different colors. Look carefully at a bright blue star and a bright red one. In August, try Vega and Antares, in December, Rigel and Betelgeuse, and in April, Arcturus and Spica. These are not like traffic lights; starlight is much more subtle. Vega is not blue but has a bluish tinge, Antares is reddish, while Arcturus has a yellow-orange hue. There are exceptions though; later we will learn of a star in the depths of space that is as red as a drop of blood.

A star's colored rays can teach us. Just by looking at a star, we can learn about its history and its type. Let's try this with the two brightest stars in Orion, Rigel and Betelgeuse. Look carefully at Rigel, the bluish star in the "lower right" or southwest corner of the constellation. Blue signifies that the star is hot, much hotter than the Sun. Betelgeuse is reddish, and much cooler than our Sun. Color is indeed a clue to temperature, in that blue stars are very hot and that red stars are much cooler. At 6000 degrees Celsius, our Sun is about average in temperature.

2.3   Distance

Are the stars sprinkled evenly throughout the heavens, or do you notice areas of sky that are more crowded than others? We are not alone in the Universe: we live in a galaxy that has some hundreds of billions of stars. While our nearest neighbors in space are scattered all over the night sky, the more distant ones tend to concentrate on a large circular area called the Milky Way. The galactic center, in the direction of Sagittarius, is so packed with stars that it forms the brightest part of the sky. It would brighten up the sky even more were it not for enormous amounts of dark matter that block out much of the light. Further out lie other galaxies, a few of which we can see as faint smudges of light.

How can we imagine these distances? Picture your home in an outlying suburb of a great city. Your neighbors' lights are scattered all around. If you look east you see the mass of light from the city center shining at you. Off to the northwest is a soft light that carries the glow from a nearby town. Other cities are too far away for their light to be seen. Our position at the outskirts of the Milky Way allows us to see it and nearby galaxies in much the same way.

If you look just below the Big Dipper, and search down just west of Arcturus and Spica, you will encounter a section of space that offers little to the unaided eye. But a telescope may reveal many fuzzy spots, each one a galaxy of billions and sometimes trillions of stars. In this "empty" spot of sky lies the thousands of galaxies of the Coma-Virgo cluster, whose gravity is so strong that it may be pulling measurably on our galaxy.

The sky is a busy place, and big enough to offer us objects, like meteors, that may be as close as 40 miles away, or distant galaxies. You can get a taste of this great offering from your back yard.

3   A word on binoculars and telescopes

Someone may have once told you that astronomy calls for large, expensive equipment. You may even have flipped through the pages of an astronomy magazine in amazement at all the fantastic technology on display there. At some point in your development as an astronomer, you may feel that a telescope will build your interest and extend the power of your observations. For now, you are probably much better off with a simple pair of binoculars, and this is true for viewing some variable stars. Because binoculars are mass-produced and sold almost everywhere, they are far less expensive than are telescopes, even those of the same size. By taking advantage of both your eyes, binoculars present the sky in an efficient, almost three-dimensional way. In variable star observations binoculars are used widely, for many of the semiregular variables are bright enough to be followed with them.

3.1   Choosing binoculars

The only problem with binoculars is that the two small telescopes that form their optical system must be precisely aligned. So many binoculars lose their adjustment with the bumps and insults of regular use. You should at least be sure to start with a good pair; test them before purchasing by pointing them to a distant building or mountain. Holding them securely, make certain that the image in one side is precisely the same as that in the other; landmarks should fall in the same place in both circular fields of view.

Soon after you get them home, you will want to turn them skyward to see the "fireflies." Yes, when I first tried to identify Vega through binoculars, all I could see was a firefly! After a few minutes, I began to realize that my lightning bug was over 26 light years, not inches, away from my eye. I was holding the binoculars with such an unsteady hand that Vega could not appear to hold her place. This problem is easy to remedy by holding the glasses against a wall or post. As you get used to the instrument, your hand will relax and then the binoculars will begin to show you the stars.

Once you are relaxed, and Vega has settled down, you will find this bright (magnitude 0) star attended by five fainter neighbors, four of which form a beautiful parallelogram. A rich star field, Lyra the Harp rewards a look through binoculars. Next, try the Dipper, and see what faint stars attend these mighty seven. Look carefully at Mizar, the middle star of the handle. The faint star nearby is called Alcor, and in earlier times these two were known as the Horse and Rider. While looking at these star fields, pay attention to the field of view your binoculars give. How much of the Dipper's bowl is visible at one time?

Under a January sky, try the region of Orion. Looking just south of the Hunter's belt, you'll see a misty spot. This is the Orion nebula, a place of many young stars. More about that later.

What size binoculars should you get? I recommend 7×50s with their wide-field lenses. The 7×50 means that the lenses enlarge objects seven times beyond what the naked eye can see, and that the lenses are 50 mm in diameter. This combination is excellent for stargazing because, for people with normal eyesight, the bundle of light rays that enters your eye will be about the same diameter as its dark-adapted pupil. This kind of binocular is the same as that used for observations at sea, and is sometimes referred to as a night glass.

Since 7 × 50 binoculars are heavy to hold, you might see nothing but the fireflies. Because the lenses are somewhat smaller, 8 × 30s or 7 × 35s are lighter. Although they gather less light, you may find them more convenient. You may also find that a tripod with binocular adapter, available at many camera stores, will steady the images you see. Most important, choose a pair that is comfortable for you, and if it has good optics, it will probably give you years of unforgettable sights both on Earth and in the sky.

Large and expensive optics are not required to enjoy astronomy. This is especially true for variable stars, and this book will show you a number of stars that you can follow with binoculars. While most of these are large red giants with semiregular periods, some are Cepheids or eclipsing binaries which guarantee regular action over a week or two.

Observing newly discovered exploding stars, or novae, puts binoculars to good use. They can also be the main instrument for a nova search during which your hunt for a star becomes the major aim of your program. A later chapter shows how to proceed with the search for novae.

3.2   Telescopes

Telescope! a word and an idea that promises something fresh and interesting. There is a time when you have finished looking out your window, finished with gazes at land and water, a time when you turn your eyes to the stars; a time when you are ready for a telescope.

Such a variety of telescopes from which to choose! In all the stores, at all the astronomy clubs, everywhere you look, there seem to be more and more types of telescopes. Which is best? Actually, there are but three basic kinds of telescopes, and it is likely that one of these three will be just fine for you. A telescope you do *not* want is the omnipresent Department Store Telescope (let's call it a DST), which is represented by the thousands of inexpensive telescopes that have come on the market recently and that are sold at the camera departments of stores. They generally do not have lens diameters greater than 60 mm, or 2.5 inches. Also, the refractor lenses are often so poor in quality that the edges have been masked off so that little glass remains to gather starlight. A pair of binoculars is usually better than a DST.

The optical systems of these telescopes are not their only problems. Almost all DSTs are so poorly mounted that they will collapse in the slightest breeze, and even when they don't fall down, the images suffer from mounts that wobble and shake.

If the funds you have are not sufficient for anything more than a DST, then they are enough for a good pair of binoculars, an instrument that is simple to use and can provide you with low power, wide field views of the sky for many years.

Refractor

This is the traditional telescope that appears to reach out the dome in comic strips, the telescope that *looks* like a telescope. It consists essentially of two sets of lenses. At the top is a large pair of lenses that is called the objective. This objective lens gathers light and bends it so that it reaches a focus at the eyepiece, which is the lens set at the other end of the tube.

Refractors can offer fine performance if (and this is a big "if") they are well made and well mounted. In the 17th century Galileo found out that the single lenses of his telescopes did not bend all wavelengths of light the same way. To correct this, achromatic lenses of crown and flint glass are used, but in some refractors the lenses are not as well color-corrected as they should be, with the result that stars of one color may consistently appear brighter than similar-magnitude stars of another color. I had that problem once, and the refractor I used could not work for estimating variable stars.

Unless you look straight through your refractor, a cumbersome procedure that will eventually strain your neck, you need a diagonal prism to bend the light so that you can observe comfortably. This prism inverts the field of view, so that the star charts do not work well.

Reflector

Reflectors use mirrors instead of lenses to gather light. In this type of telescope, light travels unaltered down the tube and is reflected by the image-forming mirror. The light then travels back up until it reaches a flat secondary mirror whose purpose is to direct the light to an eyepiece at the side of the tube.

These telescopes have been popular ever since they were first devised by Isaac Newton over three centuries ago. These reflectors are considerably less expensive than are refractors of the same size, and are widely used by experienced observers.

Catadioptric telescopes

This type of telescope has become popular as well. The most common type, the Schmidt-Cassegrain telescope, uses a correcting lens at the front as well as a mirror at the back to gather and focus light. Such telescopes use a second mirror to send the light down the tube one extra time and as a result, they are much more compact, and more expensive, than their Newtonian cousins. Moreover, since most star charts do not conform to the specially reverted right-to-left star fields that these instruments offer when used with right angle prisms for comfort, you need to turn the chart over and see its stars by shining a light through the paper to make it useful for observing variables.

Which is best? That is hard to say. If one type were clearly best, we would have only one type to look at the Universe. Newtonians lack the compactness of the Schmidt-Cassegrain, and usually a long refractor offers a finer, more detailed image of a planet. But when one considers price and overall convenience, I recommend for a first telescope, the simple to use, easily maintained reflector. Also, AAVSO star charts work well with the field orientation offered by the reflector.

Mounts

There are two basic types of telescope mountings. The simplest is known as an altazimuth, and it comes with axes to move the telescope up and down (altitude), and horizontally (azimuth). The equatorial mount differs in that these two axes are tilted to match your latitude. Thus, one axis moves the telescope in "declination" (equivalent to altitude) while the other moves it in "right ascension" (azimuth). These words describe the latitude and longitude equivalents of the sky, respectively, and a telescope so mounted can follow a star with a single motion.

If you do decide to buy a telescope, make sure that you are comfortable carrying it, setting it up, and using it. Its mount should be sturdy but not so heavy that you cannot move it.

Using your telescope

A telescope is something special, an inanimate thing that surges to life when the stars come out, and cements your friendship with the sky. Choosing the right one is serious business. Get good advice from local amateur astronomers, whose clubs may offer star parties at which you can experiment with a number of different kinds and sizes of telescopes.

When you finally have your own telescope, you will begin the wonderful process of learning to be comfortable with it. First, focus. Point at any area of the sky and turn the focus knob until the object in the field of view is as sharp as you can get it. Also, begin with the lowest power eyepiece that you have, an eyepiece that normally allows you the widest possible field. Once you have found the object of your search, you can then study it with higher power. Remember that the purpose of your first night is to get comfortable with the stars and with using your telescope.

Begin with a bright star. What other stars nearby become visible with the aid of your lenses? Next, turn to Beta Cygni, at the bottom of the Northern Cross, and visible throughout the summer and fall. With a telescope, this star's secret companion is revealed. Commonly known as Albireo, this is a double star whose two components offer a really magnificent sight. What color is the brighter star? the fainter star? If Cygnus is below the horizon, try the multiple star in the center of the Orion nebula. Known as Theta Orionis, it may reveal its four brightest components to the light grasp of your telescope.

Now find Hercules. This constellation is known for its keystone of stars, and along one side of the keystone is a strange misty spot. Try to locate the spot with binoculars, and then with your telescope. This misty spot is actually a group of many thousands of stars as they appeared some 25000 years ago. Travelling the equivalent of seven times round the earth in a single second, light from this cluster, known as Messier 13 in honor of the comet searcher who first catalogued it, takes that long to reach us. Truly, your telescope is looking into an earlier chapter of time as it looks out into space.

If any planets are up, try looking at them. How do they differ from the stars as you watch them through your telescope? How do they differ from one another?

Finally, train your telescope on the Moon. I leave this to the end because once you look at the Moon your observing eye will take 20 minutes to readapt to the sensitivity level needed for fainter objects. Fly into the craters, climb the mountains and hike into the valleys. I have spent hours doing this and have felt as close to the Moon as if I had actually landed there.

4   Learning to see

4.1   Training your eye

Learning to see is the most important skill in all visual observing, and variable star observing provides very good training in seeing. Quickly, without looking up, describe to yourself in detail the wall behind you. How are the pictures arranged? Which are colored? Where is every lamp outlet? Think of all else that is relevant.

Chances are you may have had difficulty remembering everything, but if you are alert you can train your mind to observe closely these common scenes and remember them. If you have trouble visualizing what is on the wall in your room, will you have better luck with a field of stars in space?

Let's carry this a step further. Suppose you've just walked into your friend's house, and all of a sudden the lights flash on and 15 people yell "Surprise!" Do you think you might have an easier time remembering who was there and where each was standing? Apparently, when the unusual happens, it causes your mind to snap into an increased state of sensitivity. Your mind is capable of increased observational skill when it concentrates. Your mind and eye will be very important in your career as an observer.

Seeing is an art to be developed, a beautiful capability within you that must be nurtured and cultivated. When it blossoms, the quality of your observations, whether they involve describing the wall of your room or the brightness of a star, will increase dramatically. After Sir William Herschel discovered the planet Uranus in 1781, he described how the quality of his instruments enabled him to make detailed observations at magnifications over $1000\times$. Even today such powers are practically unheard of, and the astronomers of his time questioned his ability to see at such gargantuan magnifications. Herschel replied with confidence: "Seeing is in some respects an art which must be learnt. To make a person see with such power is nearly the same as if I were asked to make him play one of Handel's fugues upon the organ. Many a night have I been practicing to see, and it would be strange if one did not acquire a certain dexterity by such constant practice."

With variable stars, your ability to see will to a great extent determine both the quality of your observations and the degree to which you can see faint stars. Always keep in mind Herschel's admonition about learning to see. It is an art which takes time and lots of practice to develop. Whenever you use a telescope, especially a small one, try to push its capabilities to the limit; not with an eyepiece, but with the astuteness of your own eye. Practice each night to see more details on the face of Jupiter. Struggle to intensify your perception of the stars around the Ring Nebula in Lyra (57 in Messier's Catalog) or the detail in the Little Dumbbell Nebula (M76). You will be astonished at how your telescope *appears* to be getting larger in size. It is not just your telescope — it is that mythical combination of observer, telescope, and sky that is working more efficiently and in greater harmony with the passing of each clear night.

4.2   VZ Camelopardalis

Now let us apply all of this to a bright but difficult variable that loves to trick you. I shall never forget the episode of the little cosmic siren that lured me into friendship and trust until I finally realized her variations were mostly due to my own lack of perception.

The "friend" is a red giant star named VZ Camelopardalis (see Fig. 28.7); a star situated so close to the north celestial pole that it can be seen in much the same part of the sky every hour of every night of the year. It has a listed variation of only 24 days between magnitudes 4.8 and 5.2, not very much, but possibly interesting. I began observing this star every few nights and quickly found out that its magnitude was indeed changing — a real variable star! But then, ever so imperceptibly, I began to notice that the star was not varying by very much and that its pattern seemed to coincide with the phases of the Moon, in a way such that when VZ was bright, the Moon was also bright. I also began to notice that whenever VZ would brighten up the sky would become hazy.

What was going on with VZ Cam? As the nights passed it became clear that VZ's period of 24 days seemed illusory. Before the observing period ended I saw a variation of only 0.1 magnitude, from 5.2 to 5.3, in VZ. In fact, in all the years since, I have not seen it go below magnitude 5.4; it is rare to see it even hit that mark and rarer yet to see it brighten to 5.1 or more.

Possibly VZ Cam does offer the variation suggested on the chart, but I have never been able to detect it visually. Only because VZ Camelopardalis can teach you about learning to see, I recommend this difficult star. By the way, after your observation, when you get inside, before you turn on the light, how many pictures *are* on the wall?

2

Getting to know the variables

5   Meeting the family

Stars are like people. Just as there are, in a sense, billions of types of people, there are tens of thousands of types of variable stars. However hard we try to classify variables, we always run into exceptions, and when we create a new category for the exceptions to the old category, that usually turns out to have exceptions too. Although it is an exaggeration to say that every variable star in the sky is in a class by itself, it is useful to think of variables as individuals capable of surprises.

There is a different aspect, however, of the philosophy of classification. Just as we can find strength and beauty by looking at the diversity of language, behavior, and culture in people, a look at the kinds of variation in stars will help to show just how extraordinary this field of study really is.

Some classifications are obvious. You would never want to confuse a stately, mature, slowly varying red giant like Mira with a little dwarf star that erupts every two months. An Orion variable, changing brightness for no apparent rhyme or reason, would not be confused with a slightly unpredictable semiregular. The eclipsing binaries, revolving together in clockwork fashion, are not the same as the intricate "breathing" of the Cepheids. Astronomers recognize these broad divisions that help us both to understand the different patterns of variation and to plan our observing programs for them.

5.1   Pulsating variables: Cepheids

Intrinsically very luminous, these stars brighten and fade with clockwork regularity. Exemplified by Delta Cephei, these are supergiant stars at least three times the mass of the Sun, and with diameters up to two-thirds that of the orbit of the Earth. It is in the extreme outer layers of Delta Cephei, just several hundred thousand kilometres below the surface, that its 5.36 day pulsation occurs. These stars actually expand and contract as they vary, although the brightest phase corresponds to the densest, or smallest, physical size. In these stars, all the hydrogen has been transformed to helium, and in that outer layer, helium becomes ionized. This layer absorbs radiation from the star's interior. In each of the approximately 700 known Cepheids, some small disturbance takes place that caused the star to compress slightly, resulting in the outer layer absorbing more radiation than necessary to retain stability, and the star expands. This means, of course, that the outer layer would absorb less energy, and the star would contract.

There are several varieties of Cepheids. Stars like Delta Cephei tend to populate the spiral arms of our galaxy, while another type, typified by W Virginis, inhabits both the arms and the large center. Their periods tend to be somewhat longer, that of W Virginis being 17.27 days. W Virginis stars tend to be one or two magnitudes fainter than the Delta Cepheids. Cepheids can have extremely short periods. RR Lyrae varies in only 0.6 day, and some stars of this type have periods of less than an hour. Even this subdivision is divided further; some RR Lyrae stars brighten much faster than they fade, while others brighten and fade at the same rates. There are also "dwarf" Cepheids called SX Ursae Majoris stars, with periods of 0.2 to 0.5 day. With almost 6000 known examples, RR Lyrae stars are quite common. A related type is the Delta Scuti, also with periods shorter than 0.3 day, but magnitude ranges so small that their variation usually cannot be detected without the use of a photoelectric photometer.

With two or three periods superimposed on each other, RV Tauri supergiants vary in much more complex fashion than other pulsating variables. The Beta Cepheids have extremely shallow ranges, in the order of 0.1 magnitude, and periods from 3 to 7 hours. This class might not be so important were it not for some prominent stars listed among its members — Spica, Alpha Lupi, and Beta Crucis. Finally, some stars have light variations which are accompanied by magnetic field changes. We call these magnetic variables, of which Alpha-2 Canum Venaticorum is a good example.

5.2   Mira stars

Also known as long period variables, Miras tend to be very slightly more massive than the Sun, but this mass is so spread out that were our Sun to be replaced with Mira, Mars would orbit just outside the star's surface!

Like the Cepheids, Mira stars physically pulsate. In addition, the pulsations possibly control an amount of carbon particles, similar to soot, that are formed in the outer layers and cause a dimming of the star's brightness.

What diversity these Miras offer! R Leonis occasionally stops for weeks as it is climbing to maximum. X Camelopardalis has a short period, and varies by 4.5 magnitudes, while Chi Cygni has been known to vary by 11 magnitudes! Some Miras brighten faster than they fade; others do the opposite. There is nothing easy about classifying Miras. In fact, recent evidence indicates that all red giants may vary at least slightly. To call the whole class of star after Mira the Wonderful is to reveal a basic truth about them all.

R Ursae Minoris behaves marginally like a Mira star. However, there are enough imprecisions in its light curve to call it a semiregular, a star with considerable differences from one period to the next. The semiregulars themselves have subdivisions. VZ Cam, for instance, has a period of some 24 days, and Alpha Orionis (Betelgeuse) slowly pulsates over 5.7 years. For some semiregulars, no trace of periodicity has so far been found.

5.3   Eruptive stars

While it is convenient to classify all variables that erupt or explode into a single group, we do create some strange bedfellows. The youngest are the Orion stars, irregular variables that are associated with the diffuse nebula. These stars are often found in groups called T associations, named after T Tauri. The variables in the Great Orion nebula are a T association.

The causes of these variations are not clear. Any star embedded in nebulosity could be obscured by differing thicknesses of gas passing in front of it. However, variation may also lie nearer to the star. Surrounding an Orion star, suggests one theory, may be clouds of tiny meteoroids. Varying thicknesses of these clouds may cause the variation. Another theory proposes that bright spots or "faculae" form occasionally on the surface of the star itself.

Anyone who has looked carefully at these stars knows that not all are the same. A related type can flare rapidly; these objects are known as UV Ceti stars and also as flare or flash variables. These stars can brighten to their maxima within a few minutes. UV Ceti once rose more than five magnitudes in less than a minute. Since UV Ceti stars are intrinsically faint, we know only of some 800 stars within 75 light years of the Sun.

The young stars erupt, and so do the old. Novae are binary stars in which the smaller member is all that is left of a star that has run out of its hydrogen or helium and collapsed. A larger, cooler star is part of the system, and transfers mass in the form of hydrogen to the small member. The hydrogen forms a layer around the smaller star, and as more and more mass gets transferred the bottom of this layer gets hotter and hotter until thermonuclear reactions suddenly occur as an explosion that we see as an increase in brightness of more than ten magnitudes within a few days.

Since old novae show small variations at minimum, it helps to watch them all from time to time. The most common novae show a rapid brightening followed by a slower decline. But there are also novae whose decline and fall is more drawn out, taking as much as six months to drop two magnitudes. Some novae are so slow that their maximum brightness is maintained for many years.

Recurrent novae like T Coronae Borealis and RS Ophiuchi are stars for which more than one major outburst has been observed. Other stars repeat their explosions over and over again every few months, like U

Geminorum and SS Cygni. Z Cam stars erupt too, only sometimes they get "stuck" on their way down to minimum.

There is even a class for stars which are not really novae but are nova-like, one member being P Cygni. A less dramatic eruptive type consists of the symbiotic stars of which Z Andromedae, with its three magnitude outburst every decade or so, is an example. These are symbiotic or combination variables in which both members of a red and blue double star system vary in brightness.

R Coronae Borealis leads a class in which the star suddenly drops by as many as eight magnitudes before beginning a slow and fitful recovery. There is an eruption, but it apparently consists of some dark material that actually blocks the star's light from reaching us.

By far the most dramatic eruption is that of a supernova, a star whose fantastic outburst costs a good fraction of its mass. For a few weeks such a star may shine with the intensity of 200 billion suns. Supernovae are divided into type I and type II stars; type I stars become brighter and decline faster than those of type II.

5.4   Eclipsing binaries

These are not intrinsically variable stars. Because two members of a binary star system happen to be in line with us, we observe a periodic drop in brightness as one member moves in front of its neighbor. In stars like Algol, eclipses take place at fixed intervals, with longer periods of maximum brightness, between 0.2 days to 30 years, between eclipses. Beta Lyrae characterizes eclipsing binaries that vary almost continuously, and W Ursae Majoris represents those binaries almost in contact with one another that have periods of less than a day.

6   Getting started with Cepheids

6.1   Delta Cephei

Just to get an idea of what variable star observing is all about, here are two active, easily found stars that we will observe informally, before we know too much. The first is Delta Cephei, an ideal star to begin with for several reasons. It is part of a bright and compact star group, it is usually around magnitude 4, and it is an active star, always offering something interesting for you to watch.

Delta Cephei's variation was discovered by John Goodricke in 1784 (see chapter 30), and it is the star for which all the Cepheid variables are named. The variation in this giant star is small but extremely regular with a period of several days. It is also typical that this star enjoys a leisurely decline to minimum that is followed by a rapid rise to maximum.

Fig. 6.1. Delta Cephei: 22h27m.3, + 58°10'; circle is 8°; Cepheid variable; range 3.5–4.4; period 5.37d.

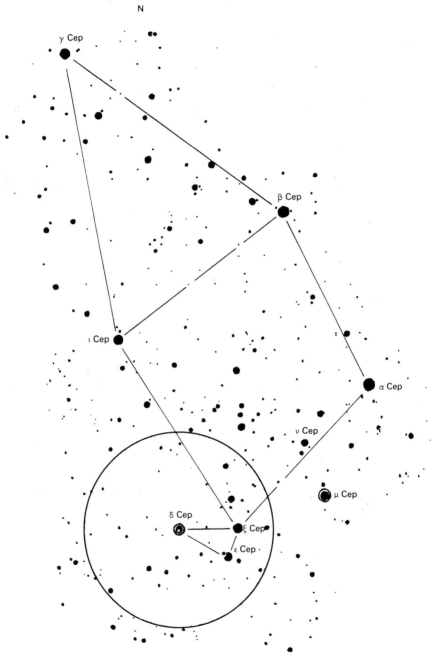

At maximum, Delta Cephei is easily visible at magnitude 3.5, and at minimum it shines at 4.4. If you observe every night or two you will soon see how it falls, then rises, week after week.

Notice the two stars Zeta and Epsilon on Fig. 6.1 and in the sky. Which is brighter? Let us give Zeta Cephei an arbitrary value of "1" and Epsilon a value of "5". Each night estimate the brightness of Delta Cephei as follows:

1 — as bright as Zeta
2 — slightly fainter than Zeta
3 — halfway between Zeta and Epsilon
4 — slightly brighter than Epsilon
5 — as faint as Epsilon

Even though we are not using actual magnitude values here we still estimate to better than 0.2 magnitude. Try this every clear night, and see what kind of variation you get. Although your first estimating session might take a while as you meander your way among the stars to the right place and then carefully make your estimate, you will soon become much more efficient.

Delta Cephei is also a double star, its 7th magnitude companion is 41 seconds of arc away. It is located in one of the exquisite star fields of the northern sky.

6.2   Eta Aquilae

A more challenging Cepheid to try is Eta Aquilae, a star whose variation was discovered by Edward Pigott, on September 10, 1784.

This star varies somewhat like Delta Cephei, with a range of just under one magnitude and a period of less than 8 days. Find Delta and Iota Aquilae on the chart and use them as comparison stars, following the same process as with Delta Cephei. Give Delta Aquilae a value of "1" and Iota Aquilae a value of "5", and record your estimate with a number, as follows:

1 — as bright as Delta Aquilae
2 — slightly fainter than Delta Aquilae
3 — halfway between Delta and Iota
4 — slightly brighter than Iota Aquilae
5 — as faint as Iota Aquilae

Make an observation every clear night and see what happens.

6.3   The Cepheids

Here is a class of stars whose historical importance has risen past the first magnitude, stars whose innocent behavior patterns have unlocked some secrets of the size of our Universe.

The story began early this century, when Harvard's Henrietta Leavitt was examining the periods of a number of Cepheid variables in what we now know are the two nearest galaxies to the Milky Way, the Large and the Small Magellanic Clouds. She noticed how orderly these stars were and how some of them had longer periods than others. When, by 1912,

Fig. 6.2. Eta Aquilae: 19h49m.9, +00°52′; circle is 10°; Cepheid variable; range 3.5–4.4; period 7.18d.

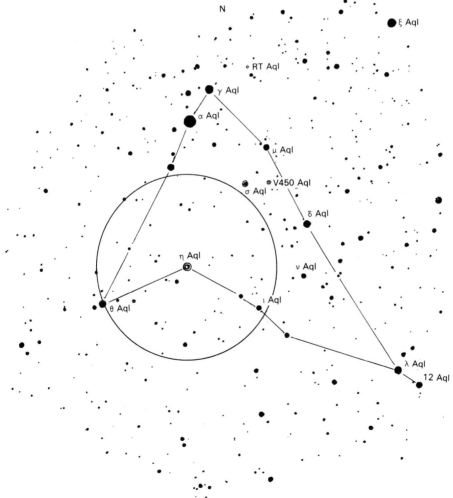

she had plotted the light curves for these variables, she then drew a graph in which the "median" magnitude, between the maximum and minimum, of each star was plotted against the period.

From these plots Leavitt noticed something intriguing, that the brighter the Cepheid's median magnitude, the longer the period. Period and luminosity were related to each other; by simply determining the period, an observer could calculate the brightness of a star. The idea of the scale was there; what was needed was the distance to a nearby Cepheid to set it. By 1917, Harlow Shapley accomplished this vital step, deducing the absolute magnitude of a nearby Cepheid. Using the period-luminosity relationship, any star in the Magellanic Clouds with the same period would have the same absolute magnitude, and the fainter apparent magnitude would be the result of greater distance. The Cepheids were the clue; Shapley used them in his astonishing discovery that the Clouds were so distant that they were not in our own galaxy, that they were galaxies by themselves.

The distance system is quite accurate for our galaxy and any galaxy in which we can find a number of Cepheid variables. Shapley used Cepheids to determine distances within our own galaxy, and Edwin Hubble later used them to learn distances to nearby galaxies. Still later, Walter Baade revised the calibration. As telescopes in space resolve more distant galaxies, locating the Cepheids within them, our concept of the size of the Universe will improve dramatically.

When you begin observing these stars, you will replay, in a sense, the research begun by Leavitt. Just as she carefully measured the brightness of stars on plate after photographic plate, so will you uncover the brightness periods and light curves for these deservedly famous objects. Stars so valuable to professional astronomers are a good target for amateur observers. Remember that their periods are short, ranging from less than a day with the RR Lyrae stars to almost two months for other Cepheids. With Delta Cephei's period of six days and one magnitude range as a reference, you can imagine how fun, though challenging, it is to observe these stars.

As with all your stars, exercise great caution in making your estimates of the Cepheids. With this type of star, competition with the professional tools is intense. Unless a professional has great confidence in visual estimates, he or she is likely to attach that light-measuring device called a photoelectric photometer to a telescope to get reliable data. Stars of small amplitude and short period lend their habits to the work of the photometer.

If your visual observations of Cepheids are to be taken seriously, they will have to be made with the greatest care. But this shouldn't prevent you from simply enjoying these stars: the need for precision should only add to the fun. Remember that as an amateur your prime purpose is to enjoy the stars, and if your work is also appreciated by a professional astronomer, that is surely a bonus.

7   Algol, the demon of autumn

While the Dipper may rule the sky of spring and early summer, different stars and new opportunities await the sky's watchers in late summer and fall.

High in the western sky is the "Summer Triangle" of Vega, Deneb, and Altair, leading their respective constellations of Lyra the Harp, Cygnus the Swan, and Aquila the Eagle. School children watch for the triangle to rise in the east during late northern hemisphere spring, for the letter "V" shape of its three bright stars promises them that summer vacation is about to begin. The position of the "V" high in the west as autumn begins is a sign that vacation is over. To the east of the Triangle flies the square of Pegasus, the winged horse. I have never thought this asterism looks much like a horse, let alone the kind that flies, but its four bright stars do resemble a baseball diamond. You can easily spot home plate and the bases, while the Milky Way to the north shines with the excitement of the fans.

A little further to the east is Andromeda the Princess, and Perseus. We now focus on Beta Persei. But why Beta? In 1603 Johann Bayer published a star atlas in which stars in the different constellations were given Greek letters in approximate order of brightness. Many stars are numbered instead, according to John Flamsteed's 1725 catalog, which used increasing numbers from west to east in each constellation.

Beta Persei, the demon star of mythology, is more commonly known as Algol, the star that winks! Every 2.9 days, or 69 hours, Algol drops in brightness by a full magnitude for about ten hours. It is one of the finest examples of a celestial event, a star that is not just there, but is visibly doing something for all to see.

7.1   Eclipsing binary stars

Algol varies because it is a binary or double star with two components that orbit each other in our line of sight, so that periodically one of its members passes in front of the other, causing an eclipse. When no eclipse is taking place, the star system is at maximum light. During the deepest point of eclipse, it is at minimum.

Almost half the stars in the sky are binary systems, containing at least two stars revolving around each other. With so many dancing partners around us, some small percentage should pass each other exactly in our line of sight, so that they appear to be getting in each other's way. In a larger percentage, the eclipse would be a partial one. In either case, when the fainter star passes in front of the brighter, we would observe a drop in total brightness. We know of over 4700 eclipsing binaries.

When the fainter companion of the Algol system passes in front of the brighter one, we see the event as an eclipse during which the brightness drops noticeably. When the brighter star eclipses the fainter one, halfway through the next cycle, photometers can record a drop a hundredth of a magnitude or so. Algol is more than an eclipsing binary; it is a multiple system since a third member orbits its center of gravity every 1.9 years.

Eclipsing binaries, as stars like Algol are properly called, are composed of stars that are much too close to each other to be seen through a telescope. Their behavior, not their appearance, gives away their secret.

Even though the eclipses take place frequently, you may have to wait before you can spot one at a convenient time. Some take place during daylight hours, or begin late at night. Check the predictions that are published each year in the *Observer's Handbook* of the Royal Astronomical Society of Canada, or each month in *Sky and Telescope* (see bibliography), for the time of the next minimum that is convenient for you.

It is probably wrong to say the eclipsing binaries are not really variable stars, that they vary only because of their geometrical alignment. If these stars are so close together that they can complete an orbit within a few days, the gravitational pull and the magnetic field interaction of one on the other is probably sufficient to induce real, though minor, change in the physical brightness of each star. However, the variation we can see with our eyes is due to their eclipses.

Most of the time, Algol shines brightly at 2nd magnitude. If you see it drop one night to magnitude 3, look carefully at this star, 82 light years away, an example of the Universe working right before your eyes.

7.2   Beta Lyrae

Let's meet Beta Lyrae, an eclipsing binary bright enough to yield its light to the unaided eye or binoculars. Like Algol, this star does not really vary. What appears to us to be changing brightness is actually the effect of a fainter star slowly sliding in front of its brighter partner. One star just keeps getting in the other's way, and your job is to find out by how much and for how long.

We shall use the same recording system as for Delta Cephei. Use nearby Gamma and Zeta Lyrae, two other stars in the delicate celestial harp of Lyra. If Beta is as bright as Gamma when you first watch, record "1"; to indicate it is faint as Zeta, use "5". Follow the above pattern for the in-between values. You will be estimating to 0.2 magnitude. If you want to estimate to 0.1 magnitude, you can use 1.5, 2.5, and so on for the intermediate stages. Make an observation every clear night and, if you have been fortunate with the weather, compute the period you have seen. I suggest that you begin your campaign when the weather is likely to be good over a few days.

Once you have observed a variable for a period of time, try drawing a light curve that shows its behavior. With time intervals, usually indicated by Julian Day, plotted along the *X*-axis, and magnitude estimates marked

Fig. 7.1. Beta Lyrae: 18h48m.2, $+33°18'$; eclipsing binary variable; range 3.3–4.4; period 12.91d; circle is 6°.

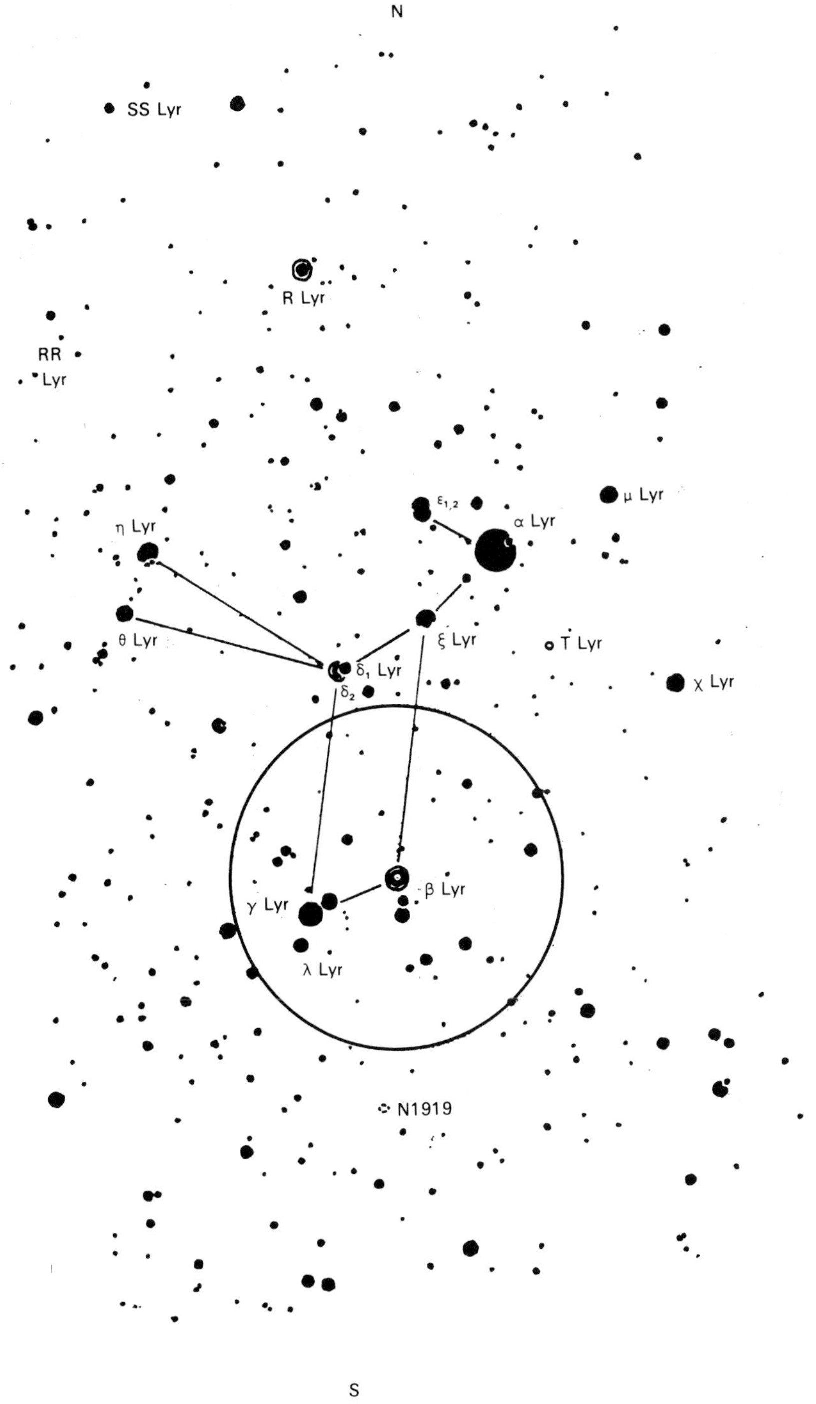

on the *Y*-axis, you can easily turn a page of numbers into an interesting portrait of the activity of a star.

Careful observations over long periods of time have shown that Beta Lyrae's period is lengthening by about two minutes a year. This system is quite fascinating. One of Beta Lyrae's members is surrounded by a disk of hydrogen, but the other star sends material into the disk as well. Also, a large, expanding, shell of gas surrounds both stars.

With experience, your observations and those of other amateurs and professionals may teach us more. You can read about other eclipsing binaries in the four chapters that look at variable stars in each quarter of the year.

8   How to estimate a variable

8.1   Z Ursae Majoris and the AAVSO method

A huge red giant, Z Ursae Majoris displays some unusual behavior in this late stage of its life. Although it usually ranges from magnitude 6.5 to 8.3, it occasionally surprises us. Once I watched it drop to 9.5 — more than a full magnitude below its normal minimum.

With the earlier starter stars we used a simple method where 1 represents the brightest and 5 the faintest. Now that we have some understanding of what the process means, we need not use it; now we are going to play the variable star game by the rules as outlined by the American Association of Variable Star Observers.

Look closely at the two charts (Figs. 8.1 and 8.2) for Z Ursae Majoris. Fig. 8.1 is designed for binoculars and has north up. Fig. 8.2 is meant for Newtonian viewing where the image is usually inverted. South is up, north down, and east and west are exchanged.

In Fig. 8.1 you recognize the Big Dipper, and near the top of the bowl is the circle-and-dot symbol for the variable star. Near other selected stars are numbers like 66, 54, and 64. These numbers represent magnitudes and tenths, but the decimal point is left out to avoid confusion with the other points that represent stars. Therefore, read "66" as "6.6" and so on.

If you look at the area with binoculars, and are able to find Z and see that it is easily visible, then you can use Fig. 8.1. Is Z brighter than magnitude 5.9 but fainter than 5.4? If not, does it fall between magnitude 5.9 and 6.4? Or between 5.9 and 6.6? Say that it falls between 5.9 and 6.4; we follow exactly the same procedure as we did for the earlier stars, but instead of using arbitrary numbers we now use magnitudes. If Z is as bright as 5.9, of course call it 5.9; if it is just slightly fainter than 5.9, then its magnitude is 6.0. Now we have to be even more careful. If it is about midway between the 5.9 and the 6.4, but ever so slightly closer to the 5.9, then the magnitude is 6.1. If it is ever so slightly closer to the 6.4 then we record 6.2. If it is almost as faint as 6.4, then the magnitude is probably 6.3.

You have just made a variable star estimate in the tradition of over four million observations sent to the AAVSO since 1911.

With patience and luck, it was easy.

Or it could have been a nightmare. You may not have found the star at all! If that is your story, you will have to go to the more difficult telescopic chart (Fig. 8.2) with south up.

Make sure that the finder of your telescope is aligned precisely with the main instrument. The way to align the finder is to observe first a bright star or planet in reverse: instead of going from finder to telescope, center the object in the telescope first, then adjust the finder so that it points to the same object. The north pole star Polaris is useful for this purpose since it hardly moves as the Earth rotates under it.

Center the finder so that the magnitude 5.9 standard is right in the middle. Then find it in the telescope, and work past the 8.8, the 9.0, and the 8.6 stars to Z. This will definitely take some time, so be patient! You will use this process, called star hopping, quite a bit in your astronomical career as you search for all kinds of objects. It teaches you the sky on many levels, and can be used with the naked eye, binoculars, and finders, and with large telescopes.

Fig. 8.1. Z Ursae Majoris. Circle is 5°. Finder chart for binoculars.

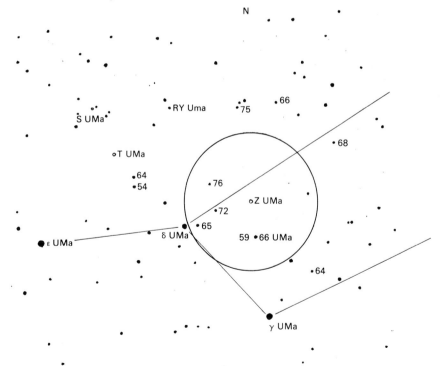

I still make the occasional error in not doing what I have just told you to. Struggling to find a star with the telescope is the single step that most often dampens the enthusiasm of a new variable star observer. Just be patient, and eventually you'll get it into view, and you will make your estimate. Having difficulty finding your star is part of the "tradition" too.

This tradition goes back as far as Herschel, and is now based on a "step" method proposed by Argelander during the nineteenth century, and refined by F. W. Pickering, who, in the AAVSO's earliest years, measured visual magnitudes to the comparison stars around variables. This work marked the beginning of the AAVSO's collection of charts, enabling an observer to estimate the variable's magnitude directly from the comparison stars. Pickering's work led to the "comparison star sequence method." Also known as the "AAVSO method," it is by far the most commonly used approach to estimating variable stars. Essentially, it is the method you just followed.

Fig. 8.2. Z Ursae Majoris. Circle is 1°. Finder chart for Newtonian telescopes.

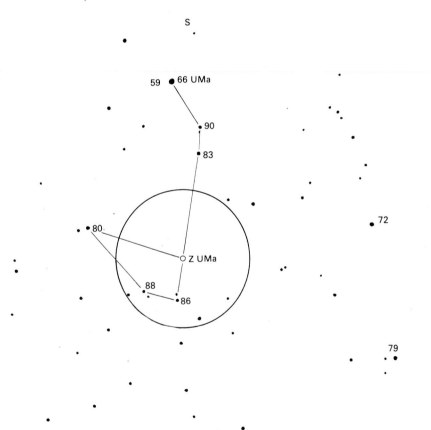

8.2   Other ways of estimating variables

Here are some of the ways through which our procedure has evolved. These ways involve some reduction after the observation, and are not normally followed by the AAVSO. I include them for your interest.

Fractional method

The "fractional" method, does not, in its first stage, need predetermined magnitudes for the comparison stars. You simply choose two stars, *a* being brighter than *b*, so that the variable is somewhere between in magnitude. If the variable were three-quarters of the way from *a* to *b*, you would record your estimate as "*a*,3,*V*,1,*b*."

Now let us reduce this observation. We find out that the brightness of *a*, the brighter star, is 8.6, and that of *b* is 9.0, with a resulting difference of 0.4 magnitude. We next divide by 4, the sum of our 3 and 1 estimate, with 0.1 as a result. From the brighter star, we calculate:

$$8.6 + (3 \times 0.1) = 8.6 + 0.3 = 8.9$$

Or, using the fainter star, we calculate:

$$9.0 - (1 \times 0.1) = 9.0 - 0.1 = 8.9$$

Herschel-Argelander method

In the "Herschel-Argelander method," we use "steps" which are defined as the smallest difference between two stars that your eye can distinguish. Before you consider the variable at all, you define steps among several comparison stars, 0 if two stars are identical in brightness, 1 if one is just slightly brighter than the other. As you become more experienced, the value of the step that you determine is likely to settle down to about 0.1 magnitude.

Pogson's step method

Norman Pogson, one of the best known variable star observers from the mid nineteenth century, developed a procedure that differs from that of Argelander in that each "step" is precisely 0.1 magnitude. It involves comparing a variable with a single comparison star using a previously memorized interval of 0.1 magnitude. You then observe the variable again, using a different comparison star. The variable's magnitude is deduced later. Your first observation might say that the variable is "*a* − 3", or three steps, or 0.3 magnitude, fainter than the brighter comparison star. Since you have already memorized what a tenth of a magnitude, or step, is, this observation is independent of the next, which considers the fainter star. In the second observation you might say "*b* + 1", meaning that the star is 1 step brighter than the comparison star "*b*". Later, we would find out that $a = 8.6$, thus $a - 3 = 8.9$. If *b* were equal to 9.0, the $b + 1$ would also equal 8.9. Remember that a star gets fainter as its magnitude number gets higher.

Pogson's mixed method

A variation close to the AAVSO procedure is "Pogson's mixed method," in which the observer knows the magnitude interval between the two comparison stars. Say that Z Ursae Majoris is in between the 8.6 and the 9.0 stars. There are four steps between the 8.6 and the 9.0 stars. In this method, if you see the variable as one step fainter than 8.6, you then independently see how many steps brighter than the 9.0 it is. If it is other than three, then one of your "step" observations needs to be redone.

The details of a method may be less important than having a clear understanding of what you need to accomplish, to identify a star correctly and to estimate its brightness with care.

9   Names and records

As the number and quality of your observations grows, your need to record them accurately and efficiently also becomes apparent. Since the recording procedure aids the accuracy of your observations, I suggest that you record your observations in a way that is designed to prevent your being influenced by previous observations.

First, record your data in an observing log that contains the results for each night's observing session. This record may well contain information about all the observing you do, including the Moon and planets, double stars and galaxies, even the Sun. You would begin with the date, times, observing site, weather conditions, seeing, quality of the sky, and instrument used. Then you would list each star that you observed during the night, by name, time, and magnitude estimate. This should be a permanently bound notebook, not a spiral or ring binder that will fall apart before it is of use to someone else.

The second step begins after your session is over, when you transfer your data to a special file for each star. If you file all of your variables on index cards and if you keep them in the order of variable star designation, you will find the compilation of your monthly observing report an easy operation. Several generations of variable star observers have found index cards or sheets of paper appropriate, but now computer programs may handle these records more quickly. Remember not to take your cards or data files outside with you, lest their listing of your last observation influence your newest observation.

At the end of each month, you should prepare a report designed to go to a formal group that can reduce it and distribute it to the professional community. The AAVSO form is reproduced opposite.

Fig. 9.1. AAVSO observation sheet. Reprinted by special permission of the American Association of Variable Star Observers, through its Director, Janet Mattei.

VARIABLE STAR OBSERVATIONS
For
THE AMERICAN ASSOCIATION OF VARIABLE STAR OBSERVERS

ASSOCIATION OF VARIABLE STAR OBSERVERS • THE AMERICAN • FOUNDED 1911 • AAVSO

Report No. . . . Sheet . . of . . .
For Month of . . . . 19 . .
Observer . . . .
Street . . . .
City . . . . State . . . . Zip . . .
Time Used, G.M.A.T., or. . . . .
Instrument . . . .

DO NOT WRITE HERE
Recd. . . . .
Pltd. . . . .
Ackd. . . . .
Posted. . . . .
Ledgd . . . .

DESIGNATION	VARIABLE	JUL.DAY&DEC.	MAGN.	DESIGNATION	VARIABLE	JUL.DAY&DEC.	MAGN.
TOTAL NUMBER OF STARS OBSERVED				TOTAL NUMBER OBSERVATIONS			

Observations should be sent to Headquarters, 187 Concord Avenue, Cambridge, Mass. 02138, as soon as possible, after the first of each month.

9.1 Designation

The AAVSO uses "Harvard Designations" derived from a star's 1900 coordinates. In the case of R Coronae Borealis, designation 154428 refers to the January 1, 1900 coordinates of 15 hours 44 minutes, +28 degrees north declination. Mira's designation of 021403 recalls its 1900 position of 02 hours 14 minutes, and 03 degrees *south* declination.

The positions of R CrB and Mira will not remain constant for two reasons. These stars do have a proper motion across the sky but over 100 years this is almost negligible. The second factor, *precession*, results from our planet's turning like a moving child's top, a slow, stately wobble that takes about 26 000 years. We see the effect of precession as a slow change of the position on the sky of the north and south celestial poles and therefore a change in the right ascension and declination of every star in the sky. Thus, the 1950 and 2000 positions differ somewhat from that of 1900.

Why do we use the old 1900 positions? The answer is based in the rise of variable star astronomy at the turn of the century. The discovery of so many new variables created a need for a logical system of designations, and the coordinate system provided it. We know that the actual positions for each star are different now, but changing the designation to keep up with it would introduce needless confusion.

9.2 Variable

In this column we name the variable star. It is true that naming the star twice may be redundant, but it does provide a self-check that the correct observation belongs to the correct variable.

Variable stars are named according to a strange combination of letters and numbers that you might see as a scientific embarrassment were you not aware of the historical reason. The system was started in the middle nineteenth century by Argelander. To avoid confusion with the lower-case a to q that Bayer's 1603 system used for naming stars, Argelander began with an upper case "R" to signify the first variable discovered in any constellation. Practically all the Rs are Miras with large amplitudes and thus rather easy to detect. The few exceptions, like R Coronae Borealis, at least share the characteristic of large variation.

The second variable found in a constellation was assigned the letter S, and the third T until the ninth variable received Z. The system entered its new generation with a tenth variable assigned a double letter RR, followed by stars known as RS, and RT, up to RZ. The system continued not with the odd title SR but with SS, then ST until SZ, then TT to TZ, and so on until ZZ.

As variables continued to make their fluctuations known, nomencla-turists made the best of an increasingly awkward situation by using upper-case double letters, as AA through AZ, BB through BZ, and on to QZ, leaving out J. Inevitably, one day one constellation produced enough variables to exhaust these 334 letter combinations. Yielding to the need for increased simplicity, astronomers decided to continue with V335 and V336. There now is a V4100 Sagittarii.

Exceptions to this complex system still remain; if a star had already been named before someone discovered its variation, the old name, like Omicron Ceti and g Herculis, would remain. Also, provisional designations of "Nova Cygni 1975" or "Nova Vulpeculae 1984 No. 2," are assigned to novae before their permanent titles are bestowed on them.

9.3   Date and time

Recording accurately when you observed a star is extremely important. If you are comfortable simply recording an observation as September 21, 1988, 22:05 EST, then do that, but be consistent. Designed by Joseph Justus Scaliger in 1582, the Julian day system is a different way of recording time that enables us to compare behavior patterns of stars over periods of years. Julian day 1 was assigned to have occurred at noon on January 1, 4713 BC. This arbitrary date happened to mark the start of three independent cycles of Sun and Moon phenomena as well as a political interval of tax collection. Although Table 9.1 lists Julian days for noon on January 1 UT for several years, I encourage you to use the highly convenient AAVSO calendar, published each year, with Julian days listed for each day.

Incidentally, the Julian day system has nothing to do with the Julian calendar. Apparently Scaliger named his system to honor his father, Julius Caesar Scaliger.

Say you have an observation of R Leonis that you made on March 1, 1967 and another observation that you made on February 24, 1981. How many days after the first observation was the second? Using months and years would make this a tiresome question to answer. We know, however, that the Julian day of the second observation was 2444660, and that the day of the first was 2439551. By simple subtraction we conclude easily that the second observation was 5109 days after the first.

Julian days are measured from noon to noon, UT. Universal time is the time used in Greenwich, England, and to convert your standard time to it you need to subtract or add a number of hours. Don't forget that the number changes if you switch to daylight time.

Once we have started using the Julian day system, why not continue by dividing each day into tenths so that each tenth occupies slightly less than an hour and a half? Use the Table 9.2 convert from Universal Time to tenths of a day. An observation made between 3:36 and 6:00 Universal time would be recorded with 0.2 being added to the Julian day, and one between 6:01 and 8:23 would have 0.3 added to the Julian day.

Table 9.1.   Julian Days

Noon, January 1 UT.	Julian day
1987	2446797.0
1988	2447162.0
1989	2447528.0
1990	2447893.0
1991	2448258.0
1992	2448623.0
1993	2448989.0
1994	2449354.0
1995	2449719.0
1996	2450084.0
1997	2450450.0
1998	2450815.0
1999	2451180.0
2000	2451545.0
2001	2451911.0
2002	2452276.0
2003	2452641.0
2004	2453006.0
2005	2453372.0
2006	2453737.0
2007	2454102.0

Table 9.2.   Decimals of day

Decimal of day	UT (from noon)
0.1	0:00–3:35
0.2	3:36–6:00
0.3	6:01–8:23
0.4	8:24–10:48
0.5	10:49–13:11
0.6	13:12–15:36
0.7	15:37–15:59
0.8	18:00–20:24
0.9	20:25–22:47
1.0	22:48–0:00

For observations requiring greater time accuracy, you can record times to two or three decimal places, or simply use hours and minutes.

9.4   Magnitude

In this last record, you note your carefully made estimate to the nearest tenth of a magnitude. Remember that an observation is not necessarily wasted if you did not see the variable. Say that you suspect that the star you have looked for in vain is about 12th magnitude but that the faintest star you can see is 11.2. Record your estimate as (11.2. If your estimate is the only one that exists, your record that the star was fainter than 11.2, would have value. If your confidence in an estimate is less than complete, add a question mark (13.6?) or colon (13.6:) to your estimate, and at the end of your report explain the reason for your uncertainty.

10   Observing hints

While variable star observing is a specialized branch of observational astronomy, the basic procedures of patience and care that apply to all observing also work with variables.

Plan your program in advance, but be flexible, since the sky often offers surprises. Choose your variable carefully. Is the star likely to be visible through your telescope, or is it obviously too faint? At the other extreme, is your star so bright that observing it is a waste of your precious telescope time?

10.1 Telescope

Telescope size

This is more of a consideration than most observers realize. In a sense, each variable star has its own best combination of telescope and eyepiece. The general rule is to use only enough power and magnification to see the variable clearly but not have it so bright that it is hard to estimate. Ideally, the variable should be about two magnitudes brighter than the faintest star you can see with your telescope. If it is much fainter than that, you will have a problem of perceiving the star, and if the variable is several magnitudes brighter, so many photons will enter your eye that its sensitivity to subtle magnitude variations will be affected.

At minimum, a star might be fair game for most telescopes smaller than 30 cm (12 inches), but as the star brightens you could use a smaller telescope. (When discussing a telescope size, I refer to the size of the mirror or objective lens.)

Do you use more than one telescope? Set them all up before a big session, and plan your program so that you use each telescope for the stars that are appropriate for it.

Mounting

Some variable star observers recommend that you use an equatorial mount for your observing because the alignment of the chart will always be the same. I have never been bothered by this, and suggest that you use a mount with which you are comfortable.

Field of view

Always use the eyepiece with the widest field. This will help you see your variable in correct proportion to its neighboring standard comparison stars. Occasionally you can spend time looking for a star field in your telescope when you are unsure just how much of the chart area appears in a single telescope field of view. Use the field circle that some of the charts offer in this book, or make a wire ring equivalent to your telescope's field size that you can physically move about your charts.

Remember that a field of one degree is approximately equal to twice the diameter of the full moon. An arc second is one-sixtieth of an arc minute, which in turn is one-sixtieth of a degree.

Magnification

When the seeing is good, increasing the telescope's magnification will apparently darken the background sky and allow you to see fainter stars.

10.2 Get to know the star field

Get acquainted with the star's surroundings, and see which stars are recommended as comparison standards. Before you make your first estimate, look carefully at the entire comparison sequence, and get a good feeling for how the stars shine relative to each other at their respective magnitudes. Do the differences make sense? Occasionally a star may not appear to you to be at its posted magnitude; perhaps the magnitude is wrong, or the star may itself may have brightened or faded. If any star does not look right to you, do not use it in your own sequence.

10.3 Normal frequency of observation

Cepheids: Once per night
Eclipsing binaries: Once every 15 minutes during eclipses

Miras: Once every two weeks

Semiregulars: Once every two weeks, or once per month

Dwarf novae: Once per night; once every 15 minutes if entering outburst

R Coronae Borealis: Once per night; once per hour at beginning of descent

Orion variables: Once every 10 or 15 minutes

10.4 A note on AAVSO charts

AAVSO star chart types are lettered "a," "ab," "b," "c," and "d." For faint stars in rich Milky Way fields, "e" and "f" charts are occasionally available. The "a" charts are intended for finding variables, and for observing when the variable is bright and comparison stars are not close to it. The "ab" charts are intermediate in scale between the the "a" and the "b" charts. Both series are plotted with north up, east to the left.

The "b" series is intended for variable stars in the 8 to 11 magnitude range. These charts are reversed, with north at the bottom and east to the right. Most of these charts are based on the German *Bonner Durchmusterung*, a star atlas prepared by Friedrich Argelander and published in 1863. Since Argelander had so much to do with the early days of variable star observing, including the development of the step method and the original proposal for naming variables after letters beginning with R, it seems satisfying to use charts that are based on his work.

The "c" charts are available for some stars and are useful for stars in the 10 to 11 magnitude range. Stars fainter than that are best observed using "d" charts. The "d" series is set up particularly well. The charts cover an area of four minutes of right ascension by one degree of declination. However, the central area, covering two minutes of right ascension and a half degree of declination, contains much fainter stars than the outer area. This way the charts are useful both for orientation and for estimating, without being overcrowded. Like the "b" series, the "c" and "d" charts all have north down and east to the right.

10.5 Factors affecting observations

Averted vision

The technique of looking "out of the corner of your eye" so that the more light sensitive rods can detect fainter objects has been well demonstrated for extended objects like galaxies, and it can work for faint variables too. If the star is hard to see, try averted vision. Remember that if you try seeing too hard with averted vision, you may develop a queezy feeling.

Purkinje effect

Red light tends to build up on your retina almost like light on photographic film. If you don't watch out for this, your estimates could be off by as much as a magnitude.

When you have this problem you are a victim of the variable star version of Murphy's law known as the Purkinje effect. If a star is reddish its light will accumulate on your retina like light on a photographic emulsion. If the sky is moonlit or hazy the effect is even more pronounced, and estimating becomes very difficult and the necessary prolonged concentration can make you dizzy.

Avoid the red build-up by using one of these approaches. First, use quick glances instead of prolonged stares. Second, move the eyepiece out of focus slightly. The subtle balls of light you now see are often easier to compare than the sharply focused points. Third, try using averted vision. This idea takes advantage of the fact that the edge of the retina is more sensitive to differences in light.

If you are in the habit of taking a long, nostalgic looks at a variable star, your estimates of any red variable may be too bright. Realizing this, some observers deliberately subtract an arbitrary, and unscientific, "fade factor" from their estimates. Instead, make your estimate using the correct techniques for red stars and then you can take a long look at your friend.

Position in the field

Your estimates will be more accurate if you bring each star, including the variable and its comparison stars, successively to the center of the field of view. Stars will appear fainter as you move them closer to the edge.

Moonlight

The light that the Moon spreads across the sky can affect your observations, especially those of bright red variables. The bright Moon increases the amount of skylight, with the result that red stars appear brighter than they would on moonless nights. Like sunlight, moonlight is scattered to give the impression of a "bluer" sky. Avoid estimating red stars during moonlit nights.

Weather

Try to keep the weather conditions consistent during your observing. Cirrus clouds are especially dangerous because you may not be aware of their presence, with a resulting error in your estimate. If cirrus clouds are present, you are better off using comparison stars that are in the variable's field of view.

Even on moonless nights, haze or high cirrus clouds can add to the stray light in the sky. Astronomers conducting photoelectric photometry at their mountaintop observatories must stop for these conditions. Your own variable star program is an exercise in visual photometry, and like your professional counterpart, you must be careful.

Light pollution

Short of moving to a dark site, there is little you can do about removing the sources of light pollution in a large city. Your problem is more severe if your observing site happens to be graced with a nearby security light. You might be forced to stop observing entirely and miss all of the objects to which this book refers. Just make the best of it by avoiding the direct glare of artificial light. A screen may work well in blocking such a light.

Do not use any light pollution reduction or other filters in estimating. In reducing different wavelengths of light, these filters change the relative brightnesses of stars.

Horizons

The closer a variable is to the horizon, the more its light will be affected by atmospheric extinction. Since stars of different colors apparently are affected differently, try to observe the star at its highest possible altitude on a particular night. Give the star a chance to get high in the sky.

Patience and comfort

The longer you look for a star, especially an easy star, the more aggravated you may become, and the longer you are likely to take to find the star. How long may it take you to find a star? Tell yourself it may take all night, and don't worry about it.

Comfort in observing is more than a luxury. If your telescope mount is about to collapse, or if your observing chair isn't high enough, or if you are cold, or if you are worried about something, your work may suffer. Variable star observing teaches you to extend your mind to the stars; relax, be comfortable, and let that happen.

Some observers take three or four deep breaths and then hold their breaths to increase their personal faint light limit by a few tenths of a magnitude. The idea is that increased oxygen to your brain increases your visual acuity for a few seconds. I have never found it to work well.

Date

Take care not to use the wrong date. Since Julian days are based on Universal time, it is possible to record an observation to the nearest minute, only to be off by 24 hours. Universal time is several hours ahead of the standard North American time zones. Remember that Julian day, beginning at Greenwich noon, is half a day off "normal" time. If you are not comfortable with the system of Julian days, don't use it. Just record date and time, and time zone. In any case, be consistent.

10.6 Seven ways to ruin a variable star estimate

1. Wind. A windy night can be more devastating even than clouds. You have set up your telescope and are struggling valiantly to keep it on its mounting. Gusts of wind can cause your eyes to tear, your eyepiece to fog, and your papers and peace of mind to blow away.

2. Cloud. Not "clouds" that simply keep you inside, but "cloud" that parks itself in front of your favorite variable. Just as it is time to go indoors, the cloud gets thinner and disappears.

3. Police. You are desperate to observe R CrB before it sets over your neighbor's window. With one hand focusing the eyepiece and the other clutching a note pad, you make your estimate just in time to be arrested for voyeurism.

4. Telescope. You and your telescope have just completed a four-hour drive to a quiet, dark site to observe faint variables, only you forgot your eyepiece.

5. Visitors. Your carefully planned observing session becomes a tour of the sky for all the people who have just arrived at your doorstep.

6. Lights. The night your neighbor has 200 people over to see his newly installed floodlights will be the same night you miss that long-awaited SS Cygni outburst.

7. Children. In a spirit of good will you have encouraged your young nephew to examine your telescope. Later that evening the hopelessly clouded field tells you that the four-year-old now knows all about eyepieces.

Enjoy yourself!

10.7 Finding the maximum: Pogson's bisected chords

Unless you observe a variable every night, which is not a good idea for Mira stars, and usually impossible from our cloud-covered world, you will probably not know exactly on which day the star reached its maximum. An ingenious way to calculate the date graphically was devised by Norman Pogson. Simply plot your observations on a graph with the dates along the X axis and your magnitude estimates on the Y axis. Next, draw lines, or chords, joining points on the ascending and descending branches of the light curve that represent times when the star's brightness reached certain arbitrary magnitudes, like 7.5, 8.0, 8.3, and 8.5. If your own observations already have, for instance, an 8.3 on the rising branch and an 8.3 on the falling branch, you are ready to join those two points; otherwise you can calculate the average of the two dates on which the star's magnitude reaches certain equal points.

For your next step, use a compass to bisect each chord to find its center. Then, draw a curve that passes through these bisection points and produce it through the top of the star's light curve. It is at the point where this curve intersects the light curve that you can fairly reliably say that the night of maximum has occurred. You may find that your observations show a slightly brighter magnitude some days before or after this point. However, because your mathematical result takes into account the entire cycle of variation, it is more accurate in setting the night of maximum in order to predict the variable's future behavior. This is a simple way to reduce your own data. Over many cycles of variation, you may even learn that a star's period is different from what has been published, or that the period is changing.

11 Stately and wonderful

11.1 R Leonis

On a bitter January night, I first watched R Leonis. A cold front had just passed through, leaving a crisp starry sky. Checking my variable star chart, I began to look for R. It was frightfully cold. After an uncomfortable 45 minute search I finally found R Leonis as it rose through the haze and smog that hugged the eastern horizon that frosty Montreal night. By this time I was so cold that even the simplest and smallest motions of the telescope were magnified into an agonizing exercise that taxed my whole being. All this for a faint magnitude 9.3 star, graced by two stellar chambermaids at 9.1 and 9.6, seemed too much. It was so cold that the telescope tube froze to its mount and I couldn't even take the poor instrument inside! Quickly, observer minus telescope moved inside for some warmth. Never had hot chocolate tasted so good!

Still outside, hundreds of light years away, shone my new variable. On that frigid night, R Leonis taught me two important lessons. One was that variable star observing can be challenging and worthwhile. The other is that to observe variables properly, one must first acquire a feeling for them, a genuine concern for what they are doing, and a will to undergo some discomfort to remain in touch with them. You may not feel this the first cold night out, but you will as you get familiar with the variable's behavior.

In following nights and years I have been able to make estimates with greater ease. Instead of 45 minutes, I now need but that many seconds to find and estimate R Leonis. And instead of R being the sole estimate of a long night, it may be one of 10, or even 25. The challenge is the same, only the degree is different.

If we are starting to observe variables in March, April, or May, the best one is undoubtedly R Leonis, whose period of just under a year is typical of Mira stars. At the opposite half of the year, Mira itself offers a stately show. These are good stars to observe because they are bright (5.9–10.1 for R Leonis, 3.4–9.3 for Mira); you can watch them move through their entire cycles even with a small telescope.

Fig. 11.1. R Leonis, AAVSO chart (a). Reprinted by special permission of the American Association of Variable Star Observers, through its Director, Janet Mattei.

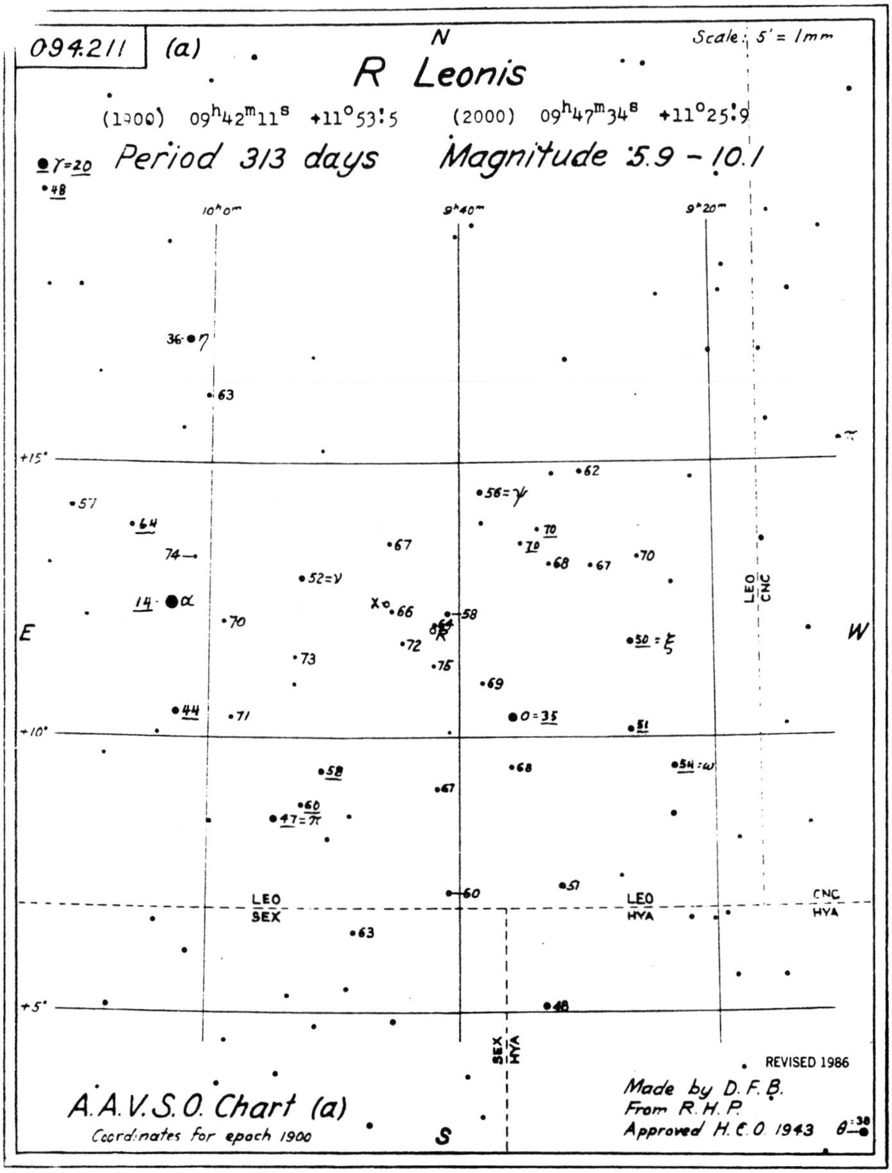

Leslie C. Peltier was the most prominent observer of R Leonis. He began his long career with a single observation of that star in 1918 (see chapter 30). The star's variation had been discovered almost two centuries earlier, by J. A. Koch, a physician from Danzig who first noticed its changing brightness in 1782. He began a string of observations that have made this variable one of the most widely observed ever. Through maxima as high

Fig. 11.2 R Leonis, AAVSO chart (b). Reprinted by special permission of the American Association of Variable Star Observers, through its Director, Janet Mattei.

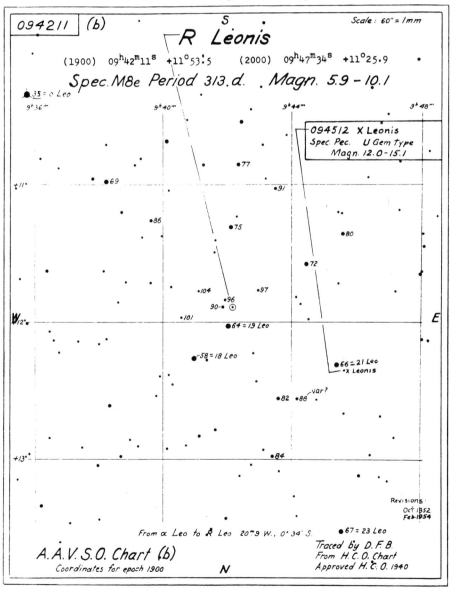

Fig. 11.3. Light curve for R Leonis. Reprinted by special permission of the American Association of Variable Star Observers, through its Director, Janet Mattei.

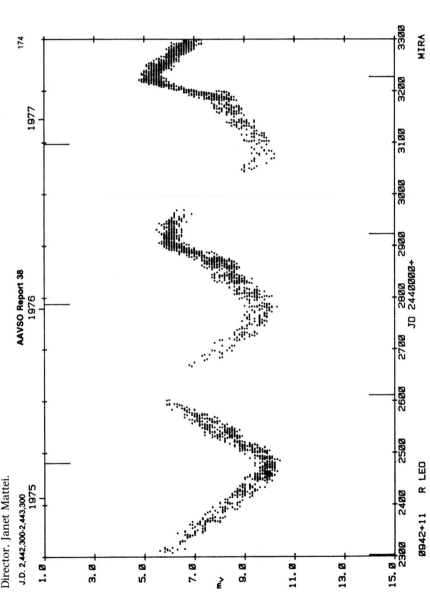

as 4.4 and minima as faint as 11.6, beginning and advanced amateurs throughout the world have enjoyed this star.

In this spirit, why not try observing R Leonis for yourself? Using the "a" chart (Fig. 11.1), work your way from Alpha Leonis toward Omicron, the magnitude 3.8 star near the chart's center. Should R Leonis be bright, you may spot it in binoculars and you might also remember that all AAVSO "a" charts are designed for such a non-inverted field. Now with the "b" chart (Fig. 11.2), find Omicron Leonis with your telescope. This "b" chart offers inverted views to match those of a Newtonian reflector, and their scales are inspired normally by the classic 1855 charts of the old *Bonner Durchmusterung* atlas.

R Leonis is a Mira star with a normal range of 5.8–10.0, and a period of 313 days (Fig. 11.3). You should estimate its brightness once every two weeks. At the end of each month, I look over my record to see what I have done. Some months the total number of estimates has been high, especially with good weather. But if the month has been clouded over either by stormy weather or by other commitments, the month's total may be low.

R Leonis helped teach me that totals are not important. As amateurs, we are interested in variables not because they provide us with a means to do more than the next observer, but to enjoy and learn from them. We are committed not to numbers but to the variables themselves and to the spirit of scientific inquiry.

R Leonis's period of just under a year is typical of Mira stars. I found that its curve is not textbook smooth during the year that began on that January night. After it had steadily risen two magnitudes from minimum, it suddenly stopped! For a whole month R just sat there, halfway between maximum and minimum, doing nothing besides adjusting itself by a few tenths of a magnitude here and there. I was sure that it would be late for its predicted May maximum. But towards the end of April, R Leonis "awakened" and hurried along, arriving at maximum only a few days after prediction! How fortunate that a distant star may appear to follow the behavior we have prescribed for it.

When you watch R Leonis, you may notice such a standstill, or the light curve may be smoother. It is uncertainties such as this that make R Leonis, a "typical" long period variable, such a thrill to watch.

11.2 Mira the Wonderful

For a few weeks every year we are graced by the light of an extra star in the sky in Cetus (Fig. 11.4). But then, just as we are getting used to the pattern in Cetus, Mira fades. Would that we could simply replace the light bulb! But nature has added Mira to her cosmic array as a beacon, an invitation for us to discover, enjoy, and learn from the world of variables. Make two estimates a month on Mira, just as astronomers have been watching her for almost 400 years.

Mira is the most famous of all the long period variables, and the first to be discovered. On August 13, 1596, fourteen years before Galileo turned his telescope skyward, David Fabricius, a Dutch pastor and an amateur astronomer, had noticed a fairly bright star in Cetus, and over the following weeks he watched it gradually fade until it disappeared. He saw it again in 1609, the year of Galileo's telescope. By 1639 a number of recorded sightings had confirmed a discovery of stunning proportions, that here was a star that was not constant in brightness. In a time when people were trying to get accustomed to the idea that the universe was more complex than had ever been known before, the discovery of Mira was especially important. In 1662 Johannes Hevelius's *Historiola Mirae Stellae* (Short Narrative of the Wonderful Star) gave the star the name it still bears with pride.

Although the charts of Mira (Figs. 11.5 and 11.6) show a certain range of brightness as well as a definite period, the figures are somewhat deceiving, since Mira does not always behave on cue. In 1799 William Herschel saw that it rivaled Aldebaran in brightness although at other maxima it may barely reach magnitude 4.0.

Discovering Mira today is not much easier than it was 400 years ago. Let's relive the uncertainty that Fabricius must have felt when he noticed

Fig. 11.4. Mira, or o Ceti: 02h16m.8, −03° 12′; circle is 10°; long period variable; range 3.4–9.3; period 332d.

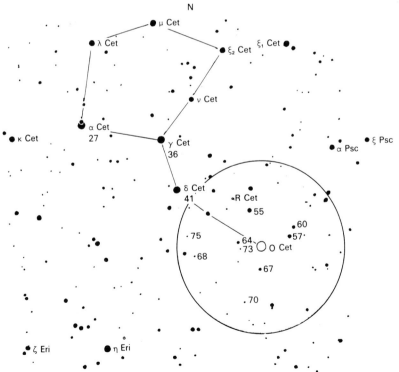

Fig. 11.5. Light curve of Mira, based on AAVSO and earlier observations from 1839 to 1970. Reprinted by special permission of the American Association of Variable Star Observers, through its Director, Janet Mattei.

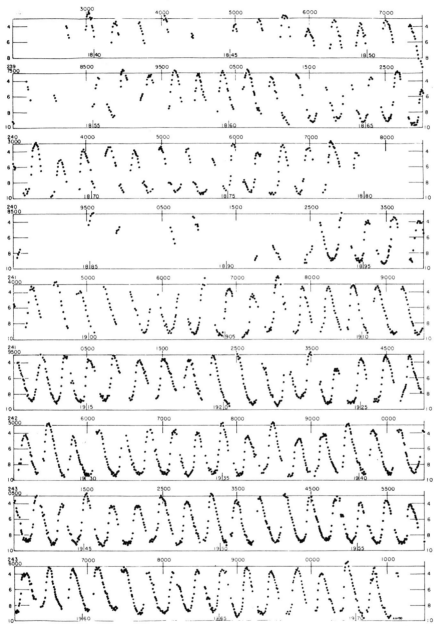

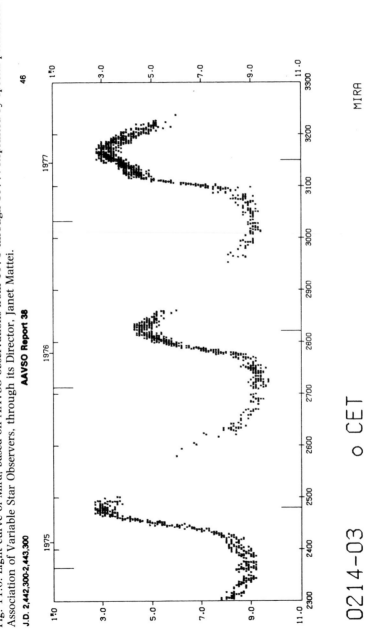

Fig. 11.6. Light curve of Mira, based on AAVSO observations from 1975 through 1977. Reprinted by special permission of the American Association of Variable Star Observers, through its Director, Janet Mattei.

Mira's strange behavior. In position, the star had been constant, but not in brightness. Was the observer at fault, or the star? Many times I have had this feeling myself when I see a variable exhibiting unexpected behavior. Cetus is a faint constellation, considerably south of Pegasus, and its bright stars are not very easy to find. Finding Mira teaches you something important in this field of observing, that once you have found it the first time, the second and third tries are much easier.

Because of their similar nature and behavior, over 5000 stars are known as Mira variables. Their prime characteristic is a distinct brightness change over several months. However, neither the amplitude nor the period of each star is precisely predictable. Mira itself, the most famous of all the long period variables, still performs with enough tricks and surprises to be well worth watching.

12 Stars of challenge

Two of the most famous Mira variables in the sky, R Leporis and Chi Cygni, are challenging, but for different reasons; R Leporis is unusually red, and Chi Cygni lies in a rich field of stars.

12.1 R Leporis

Until you've seen R Leporis at maximum, you haven't seen red. Here is a star whose redness offers us a new interpretation of color, a transcendent presence of vivid hue from a great distance.

In my early years of stargazing, I was guided by an old book by J. B. Sidgwick called *Introducing Astronomy*. During hundreds of observing sessions it taught me faithfully, pointing out the constellations one by one, as well as the inspiring contents of each. I especially remember the description of M42 — as exciting to read about as the nebula was to look at. Then I'd turn the page for Lepus, the Rabbit, just to see what glories were hidden from me in the little constellation that couldn't quite hop above the treetops of my southern horizon. It was *Introducing Astronomy* that taught me about R Leporis, Hind's Crimson Star, that shone in the sky like a drop of blood. As much as I longed to see this star, I expected I never would until either the trees fell down or I moved to a better site. Since neither prospect seemed very likely in the slowly-moving world of my youth, I relegated R Lep to a growing list of objects I would never see.

Years later and many miles away, I finally discovered R Leporis. Where was the drop of blood? It was there, but as a very faint, 11th magnitude star! Could the writer of my earliest star guide have been wrong? Or maybe my eyes were bad. Maybe I wasn't seeing red when I was supposed to be seeing red.

As that winter progressed, R Leporis slowly brightened, and at the same time its red hue deepened. When I began my second season observing R Lep, it greeted me with a blood-red, 8th magnitude glow.

R Leporis's period of 420 days is a little long for a Mira-type star. If you observe it every week or even every two weeks, you may not notice much change. Instead, try making an estimate every month. Since your look at R Leporis is a special occasion, wait for this monthly reunion until the darkest

Fig. 12.1. R Leporis, AAVSO chart (a). Reprinted by special permission of the American Association of Variable Star Observers, through its Director, Janet Mattei.

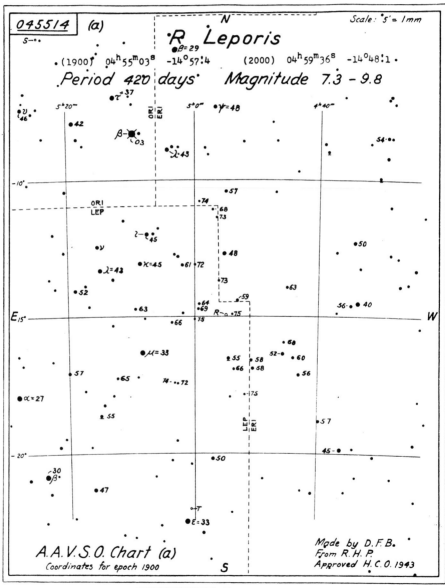

possible moonless night. This slow period also emphasizes the value of waiting, of leaving plenty of time between making estimates of all long period variables. Estimating too frequently affects you as an observer. Apart from two or three early estimates to implant the image of the star and surrounding field in your mind, the repetition of nightly observing of Mira stars could tire you. Such a program will not last. Plan your monthly

Fig. 12.2. R Leporis, AAVSO chart (b). Reprinted by special permission of the American Association of Variable Star Observers, through its Director, Janet Mattei.

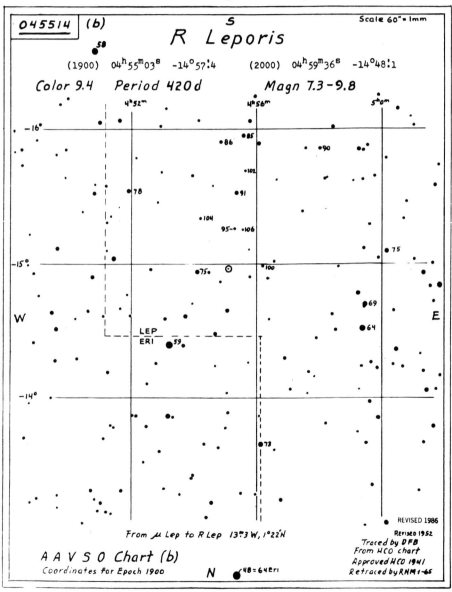

program carefully, trying to keep the red stars like R Lep in the dark half. In this manner you should see changes with every observation. Another consideration: over-sampling removes the element of surprise from your work; the memory of last night's estimate is so fresh that it prejudges tonight's estimate.

This star's brightness is not all that varies; the period itself changes from 13 to 15 months. Also, its maxima have been seen as bright as 5.6 and as dim as 7.3 and its minima have been as high as 8.1 and as low as 10.5.

R Leporis's location in the sky of January and February (Figs. 12.1 and 12.2) makes it a difficult star for an observer in a cold northern climate. It

Fig. 12.3. Chi Cygni: 19h48m.6, +32°47'; Mira variable; range 5.2–13.4; period 407d.

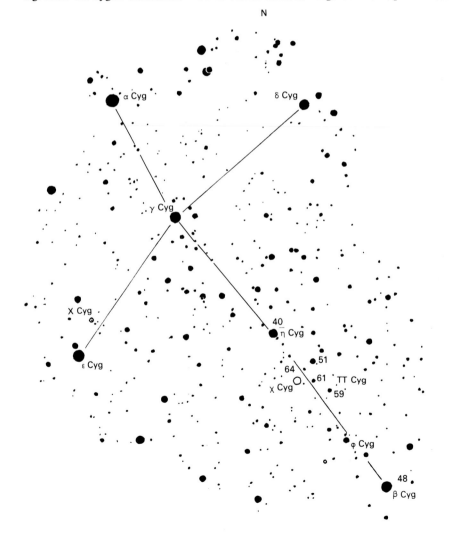

is easier to observe in balmier southern climates, where warm winter nights and a higher sky position make observing more inviting. I was stunned when I first realized how red R Leporis could get, a color of awe and mystery that beckoned further study.

You may want simply to enjoy the red color without worrying about estimating, since R Lep is really one of the great showpieces of the sky. Use an eyepiece that gives you a wide field so that you will get the widest possible field. This will result in increased color contrast between the star and its surroundings.

12.2 Chi Cygni

When Chi Cygni, designation 194632, is at maximum, it shines at a bright 5.2, adding an extra star to the long neck of Cygnus the Swan. It is often listed as a good star for beginners, which it is *if* it is near its brightest light. As Chi fades, it tends to get lost in a rich field of stars, becoming a real challenge to locate. In the 'Opening thoughts' I explained how I spent much time finding this star one August night.

Figure 12.3 is a finder chart designed to help you with Chi Cygni when it is bright. Just to the northeast, Eta Cygni is magnitude 4.0, Phi Cygni to the southwest is 4.8, and the bright star near Chi but on the other side of the line of the swan's neck is 5.1.

In Fig. 12.4, Chi Cygni tells its story during the three years from 1975 to 1977. Even in this short a time, the two maxima, or highest points of the light curve, differ by a full magnitude. Chi can do even better; its brightest recorded maximum was 3.3.

Fig. 12.4. Light curve of AAVSO observations of Chi Cygni, from 1975 through 1977. Reprinted by special permission of the American Association of Variable Star Observers, through its Director, Janet Mattei.

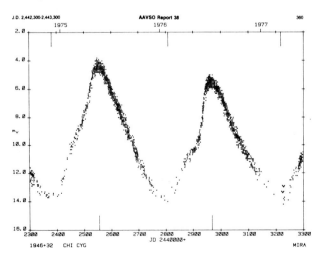

13   Bright, easy, and interesting

With growing experience you can follow several enticing stars throughout their entire range with standard 7 × 50 binoculars. These four stars are easy to find; just use the charts in Figs. 13.1, 13.2, and 13.3.

Fig. 13.1. R Scuti, AAVSO chart (a). Reprinted by special permission of the American Association of Variable Star Observers, through its Director, Janet Mattei.

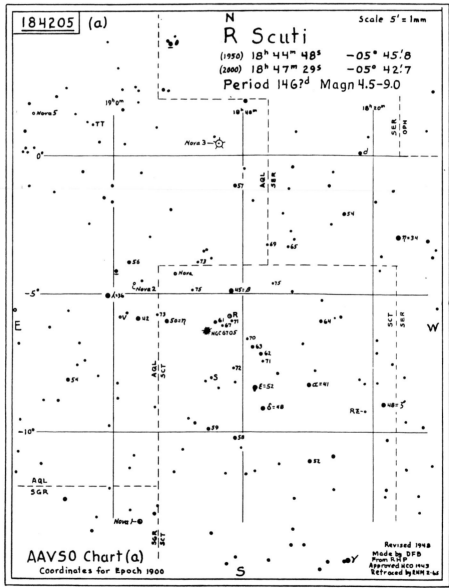

13.1 R Scuti

This most interesting bright variable ranges over about two magnitudes in almost five months. However, it shows many irregularities, and even its period, listed as 140 days, is not precise. Because the inspiring open cluster NGC6705 (M11) is nearby, R Scuti is as easy to find as it is

Fig. 13.2. R Scuti, AAVSO chart (b). Reprinted by special permission of the American Association of Variable Star Observers, through its Director, Janet Mattei.

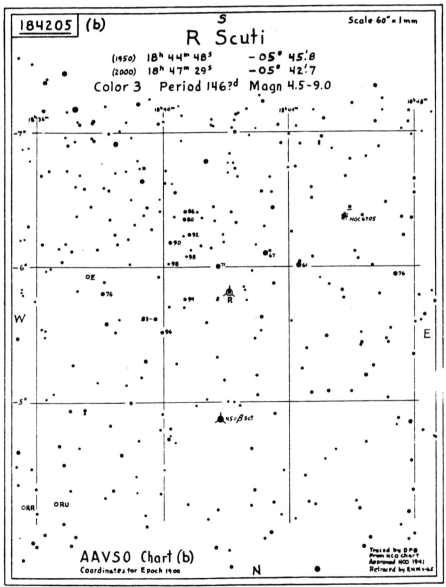

fun to estimate. The range is large, it is well placed in the sky from May to September and is a fine example of an easily observed bright variable.

R Scuti is a special type of star, in the RV Tauri class. At first its light performance mimics Beta Lyrae, although the period is much, much longer and the maxima are sharper. But then the shallower of the two minima gradually deepens until it becomes the most pronounced low point, and the star now behaves more like a long period version of Delta Cephei. You should estimate R Scuti once each week.

13.2 X Herculis, g Herculis, and RR Coronae Borealis

These are semiregular variables that provide an interesting weekly project (Fig. 13.3). X Herculis is a semiregular red giant that seems to have been deliberately placed in the sky so that you would have an easy time finding and estimating it. It is close to three bright stars, and the 6.6 and 7.5 comparison stars closely approximate the star's normal maximum of 6.3 and minimum of 7.4.

Binoculars are fine for this star as well as for its neighbor g Herculis. (The lower-case "g" not a traditional variable star designation. It follows a pattern created in 1603 by Bayer in his *Uranometria*, possibly the first atlas of the entire sky.) The star g was first seen as variable by Joseph Baxendell of England, over a century ago. Its variation is irregular, rarely showing much more than a three-quarter magnitude of change over an observing season. It is easy to locate, and is a prominent representative of the semiregular type, varying between 4.3 and 6.3.

Try the 4.9 magnitude star to the northeast and the 5.6 and 6.0 magnitude comparison standards to the west. If you make estimates every two weeks for this slowly varying star, you will have enough time to forget your previous observations so that your new estimate will be fresh and unbiased.

Now let's try RR Coronae Borealis (Fig. 13.3). Although it is not as bright as the others, you should by now have gained enough experience to observe it. Try these variables every two weeks for at least six months, and if you are in an astronomy club get other members to join you.

13.3 W Cygni

A bright red semiregular variable, W Cygni lies in the midst of a rich Milky Way star field. Usually, the 6.1 and 6.7 stars are good for estimating. The chart shows some the positions of some other interesting stars, including SS Cygni (see chapter 18), several old novae which are in the region, and a nova which appeared very close to W Cygni's position in 1978; these stars are discussed in chapter 16.

14   Betelgeuse: easy and hard

14.1   Estimating Betelgeuse

Clyde Tombaugh, who discovered Pluto in 1930, began his career with astronomy on a Kansas farm. To brighten up one long day of farming he asked himself, "How many cubic inches are there in Betelgeuse?" His answer, with what we know today, would have been 10 to the 41st power! On the next clear night he looked skyward, with a twinkle in his eye, to the reddish chief of Orion. One of its secrets given away, Betelgeuse twinkled back.

Orion is master of the winter sky. From city sky or country, from almost

Fig. 13.3. Three semiregular stars. g Herculis: 16h27m.0, +41°59'; range 4.4–6.0; period 80d. X Herculis: 16h01m.2, +47°23'; range 6.3–7.4; period 95d. RR Coronae Borealis: 15h39m.6, +38° 43'; range 7.2–8.4. Each circle is 6°.

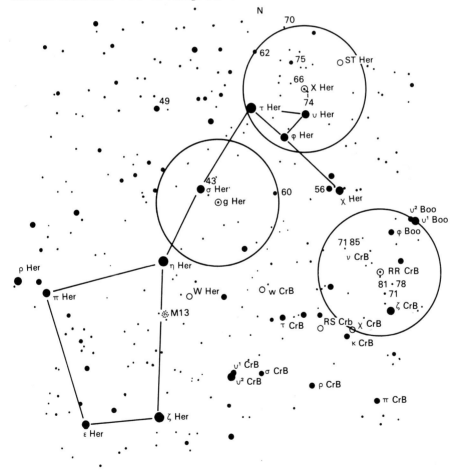

any part of the world, the majestic figure of the Hunter dominates the sky with belt, sword, and club. Look to the southeast early in a January or February evening, or to the south in the March evening, and discover Orion. The keys to this constellation are the three stars that line up in a neat row. The westernmost one is called Mintaka, a delightful Arabic name meaning Belt. Using the belt as a beacon, Betelgeuse is one of the easiest stars in the sky to find. The three stars in a row are surrounded by a four-sided figure of four bright stars. The star in the northeast corner of the figure is Betelgeuse. The best time to see Betelgeuse is on January 29, when it is in the sky most of the night.

Betelgeuse has long been a star of interest because of its brightness, its membership in the most famous of constellations, and its color. For children and the rest of us who are drawn by long names, Betelgeuse grabs our attention. Some years ago children were taught the pronunciation of "beetle juice" which naturally summoned a chorus of celestial insults from a whole generation. Imagine such a name for a star! More correctly pronounced "bet' el jews"; the name is Arabic for "armpit of the Central One." Other names for this star also implied connection to an arm:

Fig. 13.4. W Cygni: 21h34m.1, +45°13′; circle is 6°; semiregular variable; range 6.8–8.9; period 131d.

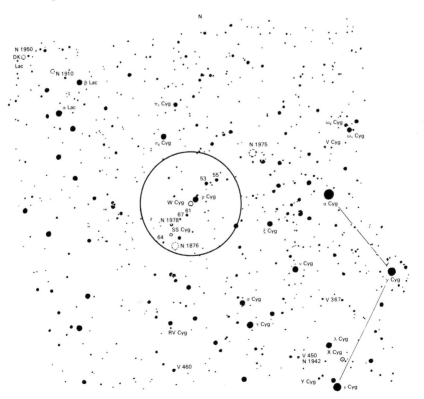

Al Dhira (the arm), *Al Yad al Yamna* (right hand). Betelgeuse was also known as "Mirzam" the Roarer who in autumn proclaimed the imminent rising of the Great Hunter.

In 1836, Sir John Herschel first noticed that Betelgeuse was not constant in brightness, and as he watched it carefully for several years, he noted its "marked and striking" variations. Betelgeuse belongs to the class of semiregular variables. It has a rough period of 2335 days (almost seven years) during which it varies by about a magnitude. If that was all, observing it would be boring. Fortunately, it has ir-regular variations that you can detect from your back yard within a few weeks.

Betelgeuse's brightness actually makes it hard to estimate, partly because it is quite red, and partly because there are not many stars in the whole sky that are bright enough to be considered as comparison stars. To estimate its brightness, we will need to choose two stars, one that approximates Betelgeuse at its maximum brightness, the other equalling Betelgeuse at its minimum. For many variable stars this would be easy, but since Betelgeuse is one of the brightest stars in the sky, finding stars to equal it at any stage of its variation is a challenge. We could try Alpha Canis Minoris, or Procyon (magnitude 0.5) in Canis Minor, which is a close approximate for Betelgeuse's 0.0 maximum, and Beta Gemi-norum, or Pollux (magnitude 1.2), which is near the brightness of Betelgeuse at its 1.3 minimum. These two stars will be our comparison standards.

To avoid problems caused by atmospheric extinction, you should use stars that are about the same altitude from the horizon as Betelgeuse. You may need to choose other standards, like Alpha Tauri (Aldebaran) at magnitude 1.1, or Alpha Aurigae (Capella) at 0.2. These stars are plotted on Fig. 1.2.

Now look closely and quickly at Procyon, Betelgeuse, and Pollux, one after another. Where does Betelgeuse fall in relation to the other two? If it is just slightly brighter than Pollux, then its magnitude is 1.2 or 1.1. If you determine its brightness to be halfway between the two standards, then its magnitude is 0.7. Since the star is not likely to change very quickly, you should estimate it no more often than once every two weeks.

14.2 Mu Cephei

Not far from Delta Cephei (Fig. 6.1) is another beautiful red star of Betelgeuse's type. Called Mu Cephei, it is so striking that the great 18th century astronomer William Herschel called it the "Garnet Star." You can use the comparison stars for nearby Delta Cephei, with which you are already familiar. Varying irregularly between 3.4 and 5.1, this star is well worth watching.

14.3 A look inside

Stars like Betelgeuse are known as supergiants because they are many times larger in size than the Sun. A blue supergiant like Rigel has an extremely hot surface. Betelgeuse and the red supergiants, on the other hand, have significantly cooler surfaces.

Since Betelgeuse is one of the closest of the red supergiants, astronomers in the early 1920s thought that, aside from the Sun, Betelgeuse would show the largest angular disk of any star in the heavens. The precise size of the disk in seconds of arc was measured by Michelson and Pease in 1922. After attaching small mirrors to the ends of two long booms extended across the front of the 2.5 m (100 inch) telescope at Mt Wilson, they pointed the telescope at Betelgeuse. Other mirrors redirected the light through the instrument and the astronomers were able to calculate that the star's diameter was about 0.04 second of arc. (Through a good 15 cm (6 inch) telescope, you may be able to see things as small as 1 second of arc.) More recent attempts using a variety of techniques yield a diameter of one and a half *billion* kilometres.

In fact, the size is so large that astronomers have used an image-enhancing technique called speckle interferometry to resolve Betelgeuse's disk. In this procedure the "speckled" appearance of a star, formed as its light is broken up as it enters Earth's turbulent atmosphere, is photographed with short exposures and then enhanced by computer. The result shows large spotted areas which could prove to be actual features on Betelgeuse. Since Betelgeuse, for all its colossal size and volume, has a total mass of only some 15 times that of the Sun, the star cannot keep up this cosmic spendthriftiness for very long, as stars' lifetimes are measured. It seems likely that Betelgeuse will undergo some drastic changes in the next few million years. One possibility (see chapter 17) is even a Type II supernova explosion.

14.4 Betelgeuse song

While you're observing Betelgeuse, you could contemplate the fate of this fascinating star by singing this song, to the tune of Richard Rodgers's *Edelweiss*:

Betelgeuse Betelgeuse
Bright red star in Orion
Soon I'm told you'll explode
So you're worth keeping my eye on.

Only two hundred parsecs away
And we know what this means

You're so near that some year
You'll blow us all to smith'reens.

Betelgeuse Betelgeuse
Speckle interferometry
Seems to show spots that glow
Spoil your spherical symm'try.

You're losing mass by convecting gas
To a stationary layer
Then there must be some dust
And an ejection sprayer.

Betelgeuse, Betelgeuse
You'll soon go supernova
When you burst I'll be first
Among those looking you over.

Matter in your circumstellar shell
Tenuous and so wide
Will in fact interact
With what's going on inside.

— Peter Jedicke, 1986

15 Not too regular

15.1 S Persei

In the field of the exquisite Double Cluster in Perseus is a special type of red supergiant, a representative of a family whose periods of variation are far less certain than those of the Miras. S Persei is a fine specimen of a semiregular variable star. The 5000 semiregular variables are a highly independent group of stars, a little too independent to predict conveniently.

Compared to other variables we've looked at, S Per can be glacially slow. If it happens to be quiescent, you may not see much change over several months. While it does have a stated period and range, these figures are somewhat decorative in that they are based on many years of observations and do not necessarily show the star's current behavior. From one month to another, it may not change at all, or it may surprise you. Observe this star at least once each 30 days; S Persei's monthly salutation yields a bonus — a simultaneous look at one of nature's finest sights, the Double Cluster, of which S is a member.

Find your way, using the chart in Figure 15.1, by going north from the cluster NGC 869 (No. 869 in Dreyer's 1888 *New General Catalog*), until you pass the 6.2 magnitude star 7 Persei, then continuing north to 8 Persei, of 6.1 magnitude. Take care with the next step, which is a jump to the northeast by about one degree.

Many of the bright variable stars we estimate belong to this same class of semiregulars. In fact, a glance through a variable star catalogue will show many such stars, with variations ranging from 5th to 7th magnitude. The reason these stars appear so bright is that they are very large

Fig. 15.1. S Persei: 02h19m.2, +58°22'; circle is 2°; semiregular variable; range 7.9–12.0; period 810d.

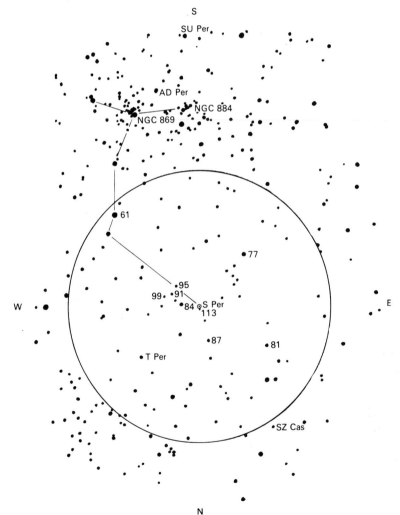

red supergiants, the term being no exaggeration, since S Persei is as much as 1000 times the Sun's diameter. We already know that bright red stars are hard to estimate. If several people try one of these stars at a time, their estimates may scatter over as much as a magnitude, which in some cases may represent the entire predicted range of the star's variation!

15.2 W Orionis

Another easily found semiregular is W Orionis, a bright red giant. The first night, just try to *find* W Orionis, using Fig. 15.2; don't attempt an estimate. It is easy to locate if you use the string of "pi" stars as a kind of celestial path that leads to the variable.

Familiarize yourself with the pattern, for nothing in that whole field should change through the years except the brightness of W Orionis. It may take two or three nights, but once you know the field well, as you know your local street pattern, *then* you're ready for your first estimate. Because it is red, take only quick glances at W Orionis and its comparison sequence. Two weeks after your first estimate, try again, doing your very best to forget your previous estimate.

Do not let memories of your first look influence your next attempt. I write everything down in one observing session log; in this way, my last estimate (how long ago was it?) is really quite hard to find. I also keep one index card that has all my estimates of W Orionis, but I don't consult it until after each estimate has been make.

Two other stars in Fig. 15.2, RX Leporis and CK Orionis, are also semiregulars worth watching. They are both subject to the Purkinje effect in this red-light district of semiregular variables.

15.3 In the mind's eye

If you are anxious enough to see your favorite semiregular star change from night to night, it probably will.

The idea behind variable star observing is to make the most accurate estimate of a star's magnitude as objectively as you can, without being burdened by emotional factors. Even constant stars may perform if you use your imagination. Although personal whim in variable star observing sounds strange, it could become a major weak link in the chain of visual estimates.

Redness in stars is not the only cause of imaginary or incorrectly perceived variations. There are several types of variables that are particularly susceptible to such problems. Eclipsing binaries are one. A psychological difficulty with eclipsers arises as you tend to observe them only when an eclipse is predicted. Knowing that an eclipse is about to take place, your mind imagines the likely drop in brightness, and if you are not careful, your expectations could actually become your recorded observations.

Fig. 15.2. W Orionis, AAVSO chart (ab). Reprinted by special permission of the American Association of Variable Star Observers, through its Director, Janet Mattei.

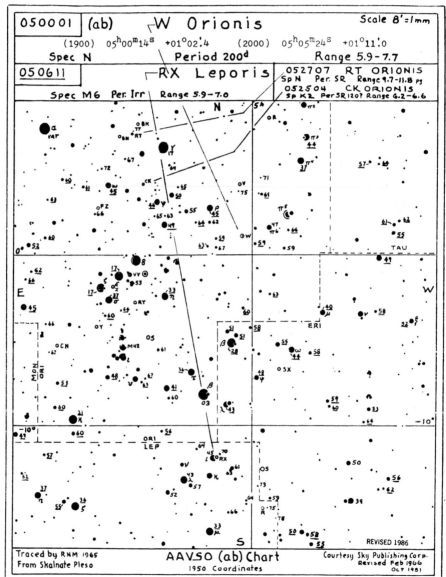

16 Nova? What nova?

The afternoon summer storm has just passed and the sky is clearing rapidly. It's a weekend evening and you decide to set up the telescope. By the time the Sun has set, your telescope is assembled and you are ready for a night with the stars. What do you do first? While the telescope is adjusting to nighttime temperatures, check out the sky. Is it just as you left it last time? A good way to begin any night of observing is to review the familiar constellations and asterisms. Slowly and methodically check each star down to 2nd or 3rd magnitude. In all likelihood, all is well, but you never know. An erupting star, a nova, might be out there waiting to be noticed as a "star out of place." A nova is not, as its Latin name implies, a new star. It is really an old star system that is exploding. In the case of an ordinary nova it is a binary system that goes into outburst as it blows off some of its atmosphere, or as a supernova, it spectacularly blows a large fraction of its mass over space.

16.1 A nova in Cygnus

Novae, though more frequent and less spectacular, are awesome in their own right. In August 1975, when I was returning with friends from an early dinner, I looked up, quite by habit, and saw what I assumed to be a slow moving satellite just north of Deneb. The others accepted my explanation and went inside, just as I noted that this particular "satellite" wasn't moving. I stared for another minute, until the truth of this new light entering my eyes began to dawn on me. It had to be a nova, a 1.6 magnitude nova in Cygnus. Later that evening I met a group of amateur astronomers at a small observatory. When I saw their telescope turned toward an area in northern Cygnus, I assumed they were examining the nova and commented on it.

"Nova? What nova?" they exclaimed, "We're looking at M39."

Not two degrees from this bright open cluster was the brightest nova in 33 years! How could they possibly have missed it?

"We found M39 using setting circles."

I am an advocate of looking up at the sky, and of finding objects when possible without the use of mechanical setting circles that are attached to many telescopes. Although they are often a great help in locating objects, I think that they deny the observer the fun of getting there. On August 30, 1975, getting to M39 without circles would have involved a process called "star hopping" with the aid of an atlas. On the way they would surely have picked up the nova.

My own preference for not using circles brought a surprise only three years later. In early September 1978 I was locating two favorite variables, SS Cygni and its apparent neighbor, a bright red semiregular called W

Cygni. SS Cygni was easy to see, but for the first time I was having difficulty identifying the field of W Cygni. At first I thought I was just tired, for I knew the field well. W Cygni was not the only red star in that field; there was a nova too!

16.2 T Coronae Borealis

This strange star is a nova whose first recorded outburst occurred in 1866, when suddenly it rose to 2nd magnitude and then slowly faded. With the strengthening popularity of variable star observing, observers naturally turned to the old novae just to see what they were doing, especially this puzzling star which seemed to vary irregularly by as much as a magnitude around its 10th magnitude minimum. Then, in February 1946, this nightly observing paid off when T Coronae erupted again, bursting overnight to 2nd magnitude.

Today T Coronae Borealis sleeps fitfully, and you should keep nightly watch on it (Fig. 16.1). Astronomers have even recorded flickering by 0.1 magnitude over several minutes. Like virtually all novae, T is a binary system, but unlike classical novae, one of the members of this system may be a large semiregular variable that causes the slow oscillations. Apparently the blue component does the rapid flickering and nova outbursts; the red one varies slowly. The orbital period is slightly less than eight months.

T Coronae Borealis is the best known example of what we call a recurrent nova. Although we suspect that all novae recur eventually, their outbursts are separated by periods as long as ten thousand years.

We know of only five recurrent novae, including T Coronae Borealis, RS Ophiuchi, T Pyxidis, V1017 Sagittarii, and U Scorpii. Their outbursts are listed in the chapters listing variables at the end of this book. The recurrences are not regular, and in the case of RS Ophiuchi can be as frequent as nine years. Observing these stars is one of the most useful programs amateurs can pursue.

16.3 Searching for novae

The careful search for novae within our own galaxy is a highly important amateur pursuit. Had such a program existed in the early 18th century, the supernova now known only by its remnants may have been discovered. The next supernova in our own Milky Way galaxy may be as bright as Venus; however, if we see it through obscuring dark matter, it may barely be visible to the unaided eye.

Hundreds of observers noticed the bright nova of 1975. Nova Cygni was the brightest nova that had appeared in over 30 years, and observing it was a lot of fun. A circular published by the International Astronomical

Union's Central Bureau for Astronomical Telegrams included people from all around the world, observing with shock the appearance of the guest star as it took its bow over Japan, the USSR, and the Wise Observatory in Israel. The report then lists observers in England who reported it as soon as darkness fell there, and then a report from the community of Rimouski, Quebec.

One observer recorded it without even being aware of what he had done. Ben Mayer of California had opened up his meteor tracking camera to record, automatically, the appearance of bright meteors. Having shown

Fig. 16.1. T Coronae Borealis: 15h57m.4, +26°04'; circle is 2°; recurrent nova; range 2.0–10.8.

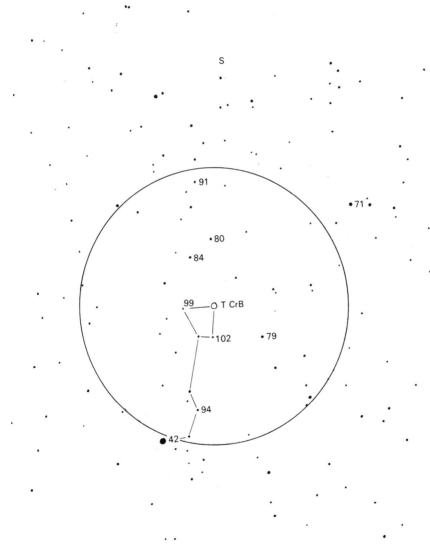

no meteors, he had thrown his photos away, only to learn at a star party of the new star. Rushing to a phone, he persuaded his son to retrieve the photos and when he returned home, he found, inscribed for all posterity, a photographic record of the rise of Nova Cygni 1975. After his experience, Mayer organized a nova patrol in which the observers photograph their selected sky areas on different nights and then compare the star fields by overlapping images projected from two slide projectors to make a device called a *Projection Blink Comparator*. This *ProBliCom* sky survey also encourages a search of the entire sky by dividing it among a group of observers.

The 1978 Cygnus nova was found independently by two members of the American Association of Variable Star Observers: Warren Morrison of Ontario, Canada, was observing a nearby variable star, SS Cygni, with a small refractor, and Peter Collins found it at almost the same time from Mt Hopkins, Arizona, using binoculars. This discovery is a tribute to what amateur observers can do with modest equipment. Open your closet and take out that small telescope that has been gathering dust instead of starlight all these years, and see what it can find for you.

The AAVSO sponsors a Nova Search Program, in which searchers have divided the nova-prone Milky Way regions into over 100 areas, most of which are 10 degrees on each side. The search is quite basic. With binoculars or a small monocular, check each star in your area down to a certain limiting magnitude. I usually check down to 6th magnitude, although some observers go down to 7th or 8th.

One aspect of the AAVSO program that I find appealing is that once an area is assigned to you it figuratively becomes your piece of celestial property. It must be tended like a garden. Make sure that the stars which populate it each night belong there, with no intruders present. Get to know your area well; one day it may reward you.

Searching for novae can be one of the most interesting aspects of variable star work, partly because it involves changing the rules of observing. In regular variable star work, as in most types of amateur observation, you choose the star you wish to see, and assuming a clear and unobstructed night, the sky will cooperate. In nova searching, the rules of the celestial game are reversed. Rather than choose a star, you choose a broad area to search. What you see in the next hour or so of searching will not be up to you, but up to the sky! You may find some fields of variables with which you are familiar; you greet them like old friends passing along a street, and then bid farewell, for you still have a distance to go. Some fields may even offer to your binocular fields some deep sky objects, and since your search area is defined by the sky as the vicinity of the Milky Way, you will likely get to meet lots of those. What will the next field bring? a bright open cluster? a diffuse nebula? an interesting asterism? a nova? When you begin, you really do not know. It is up to the sky to show you.

Selecting areas

One way to hunt for novae is through a cooperative search program in which the sky is divided among many observers. In the other method, you divide the sky yourself, creating your own patterns, or miniature constellations, out of the thousands of little asterisms that make up the starry sky. Why not begin with an obvious pattern, like Brocchi's Cluster, also known as the Coathanger, a group of stars in that familiar shape. Memorizing this pattern is simple, and if you concentrate on it each night, you may discover a guest star in its region some night, just as G. E. D. Alcock did in October 1976.

Begin your program slowly, creating only a single pattern or two each night. Eventually your search area will grow to considerable size, and you'll be surprised at how much of the sky you have memorized to 6th, 7th, or even 8th magnitude.

Keep on observing the areas you have learned, for you must recall another rule about playing the observing game this way: the sky will not wait for you. If you want to discover a nova you must be there when it happens. You need to search the sky every clear night, moonlight or no. There are exceptions, but usually the sky rewards the patient observer whose painstaking search has lasted for years, the skywatcher who goes to great lengths to observe, and who never gives up.

Discovery

You have seen a star you cannot identify. Then what? How well do you know that part of the sky? If this is one of the first times you have observed that region, check on at least two star atlases that include stars down to the magnitude of your suspect. Then you should wait at least a night to see if the interloper is not a passing asteroid. A check with a larger telescope and high power may show an asteroid's motion the same night.

Make sure that you do indeed have the correct field. The sky is funny that way; it may just offer enough stars to make you think you have the right field. (Some people have misidentified something as obvious as the Southern Cross. There is indeed a nearby "False Cross" that these observers must bear.)

If your suspect has appeared in an area with which you are thoroughly familiar, then you can proceed with much greater confidence. In any case, get someone else, an experienced observer of good reputation, to confirm your sighting.

Once you are certain, then you should notify by telegram the AAVSO, at 25 Birch Street, Cambridge, Massachusetts 02138, USA. Include the date and time of your discovery, the position of your suspect in right ascension and declination as well of the epoch of the star atlas you use (usually 1950.0 or 2000.0), the magnitude, and your name and where you can be reached by telephone and mail. With a team of experienced observers, the AAVSO can confirm your nova suspect.

Once you have discussed the event with them, and they agree that your suspect is most likely a nova, you should then inform the Central Bureau for Astronomical Telegrams of the International Astronomical Union. The Central Bureau recommends that you send your telegram using their TWX number: 710–320–6842 (answerback ASTROGRAM CAM), even if you use non-TWX services like ordinary commercial telegram companies. Add the address "Central Bureau for Astron, Cambridge Mass" but do not use the "Smithsonian Astrophysical Observatory" in the address as this causes delay. The Bureau has done superb work for more than a century in monitoring the sky's transient phenomena.

Naming of novae

A newly discovered nova is assigned the name of the constellation in which it resides, followed by the year. The bright nova in Cygnus that so many people found in August 1975 was assigned the designation "Nova Cygni 1975" and then the standard Harvard designation 210847. Shortly thereafter it was assigned an official name, V1500 Cygni. If two novae are found in the same constellation in one year, the second gets a "No. 2" after it; thus, Peter Collins in 1984 discovered "Nova Vulpeculae 1984 No. 2".

16.4 Watching an old nova

The sky has several interesting old novae for you to check from time to time. Sometimes these stars may climb rapidly by as much as a magnitude or more, and then return more slowly to their usual state. 184300 V603 Aquilae is an old nova whose 1918 eruption astonished hundreds of observers who had watched a total eclipse of the Sun just a few hours earlier. Try observing it occasionally as it hovers around 11th magnitude.

Few astronomical events can cause the sensation of the discovery of a nova. These events show us dramatically that our galaxy is full of change and evolution, and that this change can occur before our eyes. It is proof that our galaxy still has new things to show us. In a more esoteric sense, it also shows that when we are among the first to see an event, on its opening night, we can become, in a sense, a part of it. That the Universe is unfolding is a beautiful thought, but often irrelevant to our daily lives. When the Universe invites us to watch part of that process, its growth and change are driven home.

Fig. 16.2. V603 Aquilae: 18h46m.4, +00°32′; circle is 2°; nova (1918); range −1.4–12.0.

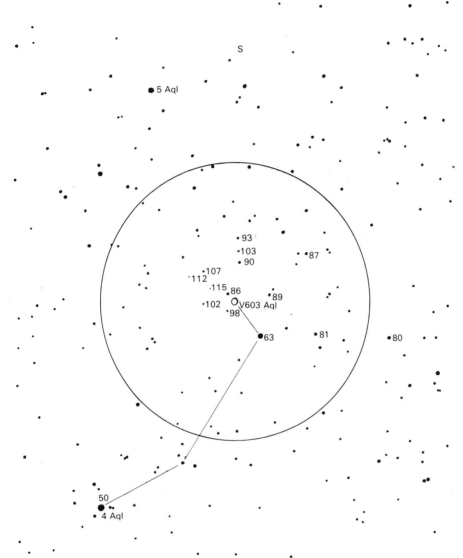

17 Supernovae

17.1 A thousand years ago

There is something magical about a star that, in its final phase of life, announces its end to the entire Universe. On July 4, 1054, Chinese skywatchers were stunned by something new in the sky. The "guest star" they reported was the outburst of a supernova, triggered by a major instability and collapse within the star. Since then at least three other supernovae have been easily visible from Earth, in 1572 and 1604, a time when European civilization was just about ready to accept new thoughts on the stability and arrangement of stars in space, and in 1987, when our understanding of the process of a supernova was finally good enough to be tested.

In 1054, native American records consisting of paintings and carvings on rocks welcomed a bright new star in Taurus. More accurate Chinese records have given us information so that we can pinpoint the date of the event as July 4, 1054. Much brighter than nearby Aldebaran, the exploding star could be visible in daylight.

What has become of that star? Not too much was known, or cared, before Charles Messier turned his telescope in that direction while hunting for comets and discovered a fuzzy looking object. It looked a bit like a comet, and Messier thought he had discovered a new one. However, when the fuzz refused to move like a comet would, he decided to prepare a list of comet masqueraders, and this object would be number one on his list. Other observers noted this object, now called M (for Messier) 1, and gradually we became aware that M1 was in the same position as the old supernova of 1054 and must have some relationship to it. We now know it to be the remnant of SN 1054, a huge display of gas marking the spot of a star's death. In 1850, Lord Rosse compared the appearance of this object to that of a crab, and the name Crab Nebula has lasted ever since.

The Crab Nebula is one of the most remarkable objects in the sky. When in 1967, Anthony Hewish and Jocelyn Bell detected signals coming at regular intervals, they wondered about their regularity, which was so precise that the scientists even joked about artificial origin, calling them informally LGMs, for little green men. Hewish and Bell concluded that these pulsating objects were the neutron stars that are the collapsed cores of stars that became supernovae. The best known of these is in the Crab Nebula, and a few years after this discovery, astronomers at the University of Arizona, using a 36 inch (0.9m) reflector on Kitt Peak, found the first visible pulsar, right at the center of the Crab. It is an object consisting of neutrons, a neutron star spinning wildly 30 times a second. This period is constant, except for a very gradual slowing down, and occasional temporary changes that may be produced by "starquakes." Although the pulsar is visible through a very large telescope, its variation is not visible without electronic aid.

Four other supernova events within our galaxy have been identified. In 1006 a star in Lupus brightened to possibly magnitude -1.0 Tycho's star of 1572, in Cassiopeia, was as bright as Venus; and Kepler's star of 1604, in Ophiuchus, was brighter than Jupiter. Early in the 1700s, a supernova erupted again in Cassiopeia. Although it was not observed at the time, its remnant survives today as a strong source that is "visible" in radio telescopes, and as a nebula of material that is expanding. It seems inconceivable that a star that brightened enough to outshine the rest of the galaxy was not discovered, but its rays were blocked by clouds of dark material between it and us, and at its maximum Cassiopeia may not have been conveniently visible.

17.2 Type I and Type II

A supernova is a truly catastrophic event, and marks the end of a star's lifetime struggle to keep its forces in equilibrium. In the 1930s, Fritz Zwicky and Walter Baade developed several models of supernova events. Most astronomers recognize the first two and refer to them now as Type I and Type II supernovae.

The story of a Type I supernova begins in the internal battle between the force of gravity which pulls toward the center, and that of its thermonuclear fires pushing the matter outward. Throughout much of a star's life, the contest is resolved as a draw. After the star has used up its supply of hydrogen, its thermonuclear fires weaken and gravity gets the upper hand. The huge core begins to contract, and what hydrogen is left ignites in a shell around the core. Thus, the star swells into a giant and then begins the process of burning its helium. Like the hydrogen before it, the helium is burned first in its core and later in the shell. As the helium supply declines, the still contracting core is left with carbon and oxygen. The star is now a white dwarf, its atoms almost touching each other, and resisting any more contraction. We call matter in this state "degenerate matter." When the Sun completes this process, it would shine weakly as a white dwarf until it cools enough to stop shining altogether.

If the dwarf is a member of a double star system, and if the other member is close by, the dwarf's gravity will start pulling matter from its companion, getting more massive. In some stars the excess matter is blown off routinely. When these explosions take place in periods of thousands or tens of thousands of years, we call them novae. If they happen in periods of tens or hundreds of years, we call them recurrent novae, and when we see the explosions every few months, we call them dwarf novae. In all these cases, by one process or another, the material that a white dwarf captures from its neighbor is successfully ignited and blown away, and the equilibrium is more or less maintained.

In some stubborn cases, the captured matter does not ignite at all and the white dwarf simply gets more and more massive. Over hundreds of thousands of years, no ignition of captured material occurs. Eventually the gravity becomes so powerful that even the degenerate matter contracts. This limit of stability, the greatest mass a dwarf can have and still stay a dwarf, was first theorized by Subrahman Chandrasekhar. When a star reaches Chandrasekhar's limit, it has reached the end of its road. The resulting fatal collapse takes less than a second, and then all the degenerate matter explodes.

Type II explosions involve very massive stars that live out their lives in a hurry, burning themselves out in perhaps just a few million years. Such a star spends most of this time fusing its hydrogen, after which it fuses its helium. As the helium supply declines, the still contracting core is left with carbon and oxygen, and in the core the temperature rises to the point where the carbon ignites. In less massive stars, say three or four times the mass of the Sun, all the carbon ignites at once, and the resulting supernova explosion blows the core apart.

What if the star is so massive, say nine or ten times the mass of the Sun, that its already very hot core causes the carbon to ignite and burn gradually? Heavier elements like phosphorus, aluminum, and sulfur generate in shorter and shorter intervals, until silicon is formed. Then, it takes just one day for the silicon to fuse into iron.

Iron simply does not release energy by burning. Instead, it absorbs energy, getting hotter and hotter. The process of nucleosynthesis, in which all the other elements were created, stops dead. In less than a second the core crashes in on itself, bringing with it enormous amounts of still unused fuel; electrons are forced into their atoms' nuclei, forming neutrons and neutrinos. In the resulting explosion, the star blows away its outer layers, and can outshine its entire galaxy. In its last gift to the Universe, even heavier elements are formed during the explosion.

What is usually left is a neutron star, although more massive stars would have cores so dense that they would crush even the neutrons so that even light could not escape. Such an extremely dense object would be known as a black hole.

17.3 The supernova of 1987

Early in February of 1987, astronomers were still living with the fact that the last naked eye supernova happened in 1604. There were two other candidates, the first in our galaxy, the other in Andromeda Galaxy. We have no evidence that Cassiopeia A, as the first is known, was observed by anyone. S Andromedae exploded near the center of M31 in 1885, and although its maximum brightness reached about 5.8, the surrounding light from the galaxy probably prevented anyone from seeing it without optical aid.

Ian Shelton, a Canadian astronomer at the University of Toronto's telescope at Las Campanas in Chile, began a patrol program of both the Large and Small Magellanic Clouds on the morning of February 22, 1987, in an effort to discover new variable stars and novae. On the morning of

February 23, Shelton developed his second plate of the Large Magellanic Cloud. The third plate was taken and processed on the morning of February 24. All plates showed a familiar feature, the Tarantula Nebula which is a giant version of our galaxy's Orion Nebula. The plate of the 24th, however, had something extra. At first Shelton thought it was a defect, but when he went outside to look at the cloud, he saw a new bright star!

At a neighboring telescope, Oscar Duhalde was checking the sky during his duties as night assistant when he too discovered the intruder. The third discoverer was Albert Jones, a visual observer from Nelson, New Zealand. At Australia's Siding Spring Observatory, Robert McNaught began a series of visual estimates of SN 1987A's brightness. During the first hectic month, McNaught's estimates were published worldwide almost daily by the *Circulars* of the International Astronomical Union's Central Bureau for Astronomical Telegrams.

Having just completed the final draft of this book, I very much wanted to see this new variable star, and having just edited an earlier draft, Steve Edberg, a well known comet scientist, wanted to see it too. Our friendship had grown from our love of comets, and we called our trip to Acapulco a comet expedition to see a supernova. After two cloud-plagued nights, the south Mexico sky cleared and we saw a bright red star just 3 degrees above the southern horizon and 160,000 light years away. We were looking at one of the major events the Universe has to offer, and since the Large Magellanic Cloud is the next galaxy to ours, it was in our back yard.

As we gazed silently at the supernova, we thought how it is a beacon of life. All life forms on Earth are united by their base in carbon. We know of only one mechanism through which carbon can be spread throughout the Universe to become part of the clouds that eventually compress to become stars and planets. That process is a supernova explosion. At the moment a star's core collapses, the heavier elements, including carbon, are forced into space. We felt a kinship with 1987A, and we were moved to be able to see it and to appreciate its significance.

17.4 Long journey of the neutrinos

You really should be quite at home with neutrinos. After all, since you began reading section 17.4, about 100 billion of them raced through your body. The trouble with these neutral atomic particles is that they have practically no mass and do not react with other particles, and thus are very difficult to detect. Occasionally, they do react with chlorine, but unfortunately, cosmic rays also react with chlorine. How does one separate the events?

The answer lies in the neutrino's ability to travel through almost anything. Cosmic rays do not travel through the Earth; if the detector is buried underground, only a neutrino would react with chlorine, producing argon, an atom which is radioactive and which can be detected. Only seventeen years before SN 1987A, the first neutrino trap went into operation. Its purpose was to detect neutrinos sent from the Sun.

These elusive particles come from other stars too, and astronomers have theorized that the moment a collapsing stellar core forms a neutron star, huge numbers of little neutrinos spread through the Universe. We know that this process occurred in SN 1987A, because some of these neutrinos actually reached our detectors.

Buried deep in a mine, Japan's Kamiokande detector observed 11 neutrino bursts in 13 seconds just 3 hours before the supernova's light reached us, and the Irvine-Michigan-Brookhaven (IMB) detector buried under Lake Erie noted 8 neutrinos in 6 seconds a few seconds later. Meanwhile, the Mount Blanc Neutrino Observatory recorded 5 neutrino pulses in 7 seconds, and the detector at Mt Elbus also picked up pulses. These neutrinos have travelled all the way from the collapsing core of the distant supernova; this is the first time ever that we have detected these particles from an event taking place in another galaxy.

The neutrinos arrived here with a vengeance. The few events that were recorded mean that, since these particles are so difficult to detect, huge numbers of neutrinos from SN 1987A actually passed through the Earth and its inhabitants, including you. Although their journey began at the moment of core collapse, they did not get too far at first, since the core was so dense that they could not escape it. When the shock wave from the explosion reached the neutrinos, it sent them cascading into space.

17.5 Observing and searching

The two types of supernovae behave differently. Although both types show very rapid rises in brightness, Type I supernovae drop precipitously, while Type II stars decline more slowly, and may rise briefly again three weeks after their maxima. In fact, SN 1987A remained near maximum light for months. The second type does not reach quite the absolute brightness of the first.

No supernova has been observed in our own galaxy since 1604. However, these events are seen regularly in other galaxies, and diligent searches by amateurs have yielded many of them. Rev. Robert Evans of New South Wales, Australia, is the leader in this search. His systematic, well-organized patrol has yielded 15 discoveries from 1981 to early 1987. Although he lost a chance to discover 1987A because of clouds, he did find 1987B. Using at first a 10 inch (25 cm) reflector made by himself and later a 16 inch (40 cm) reflector, he searches through many galaxies each clear night.

Although supernovae occur in irregular, spiral and elliptical galaxies, they are easier to spot in galaxies which we see face-on, rather than edge-on. Compare the galaxies you see through a telescope, or through photography, with a published photograph or chart. The photograph should not have been taken a telescope so large that it shows hundreds of stars you could not possibly see through your own telescope. A useful guide to supernova searching has been written by Gregg Thompson and James Bryan, and a convenient source of galaxy photographs is Juhani Salmi's *Check a Possible Supernova*. Both are described in the bibliography. As with nova searching, the first important step is to become familiar with the galaxies you search. Some galaxies have bright regions that are often confused for stars, and you can mistake foreground stars for possible distant supernovae.

17.6 Supernova song

In the utter devastation that we see as a supernova event, carbon is released into the interstellar medium of space to become, perhaps, the essence of a new stellar system and the life therein. Try considering this concept through these words, sung to the old popular tune of "When Johnny Comes Marching Home:"

> The stars go nova one by one, kaboom, kaboom!
> Nucleosynthesis is done, kaboom, kaboom!
> The supernovae dissipate
> What fusion energy helped create
> And the stars go nova in the galaxy.
>
> The heavy elements are born, kaboom, kaboom!
> And from the stellar cores are torn, kaboom, kaboom...
> Shells of gas are strewn through space,
> Distributing matter all over the place,
> And the spiral arms are littered with debris.
>
> As years go by the remnants spread, kaboom, kaboom!
> But the Universe is far from dead, kaboom, kaboom!
> To eliminate the tedium,
> The interstellar medium
> Forms the molecules that make up you and me.
>
> — Peter Jedicke, 1981

18   Three stars for all seasons

I owe Rik Hill ten dollars.

It all started one evening when we were discussing the unpredictability of these strange stars we call recurrent novae. Suddenly Rik proposed a wager. Suppose, for example, T Corona Borealis erupts every 80 years. With its last outburst in, say, 1946, I could bet ten dollars that it would next erupt around 2026.

This would be quite a wager, with uncertainty so great that were Rik to bet on 2000 and me to say 2050, we would stand an equal chance of winning.

Since I didn't want to keep ten dollars in some distant bank that long, I suggested we try a *dwarf* nova, a star that erupts about every 50 *days*, not years. "No, SS Cyg is too regular," Rik countered. "There are variable stars that make much better bets."

Anyway, I talked him into a $10 wager on SS Cygni. A week later, at 11:30 p.m., in the midst of a dark southern Arizona night, the telephone rang ominously.

It was Rik. "Have you seen SS Cyg tonight?" SS Cygni is regular? Who says? It went into outburst seven days early! And I lost my bet!

Even at minimum this star shines at a not-too-faint twelfth magnitude, visible with a city-bound 20 cm (8 inch) telescope. The field is not easy to find, but once you succeed, it is easy to remember (Fig. 18.1). Since you should watch SS Cygni's machinations every night (or several times a night if the star is going into outburst) you would lock it into your mind quickly. After you see its first outburst, I'll bet (again) you will be as hooked on it as I am.

I should have known that SS Cygni would cost me money. In October 1979 I watched it rise slowly to magnitude 11.3 and then hover there, as if it couldn't decide what to do next. Dawn put a stop to my watch, and as the day wore on I got increasingly nervous. Already I had alerted the AAVSO that an outburst might be imminent, and only two weeks earlier I had received my special personalized license plates with the letters "SS CYG" on it. What if I were wrong? Surely Arizona would recall my license plate.

With the coming of night and with fingers trembling I focused the eyepiece of the telescope. A magnitude 9.0 star was smiling in my field of view, brightening by the hour!

18.1 A look at SS Cygni

What is this strange variable? It is a dwarf nova, one of about 250 such stars that remain at a minimum for weeks and then suddenly climb. SS Cygni climbs almost four magnitudes, stays at its 8.2 or 8.3 maximum for a few days and then declines more slowly. One night it could shine at its usual 12th magnitude, but the following night it may easily startle you at magnitude 8.3, one of the brightest stars in your telescope's field of view. What happened? Since you observed the star the first night, SS Cygni went into outburst and within perhaps a few hours, brightened by almost four magnitudes. This is a dwarf nova whose bimonthly explosions attract dozens of variable star observers. The suspense increases when you realize that although the star's period is roughly 50 days, outbursts can range from two weeks to four months (Fig. 18.2). When will SS Cygni go off again? Tomorrow? Next week? In three months?

Cataclysmic variables like SS Cygni are binary systems that consist of a small star called a white dwarf and a larger, cool star. The distance between the two components may not be much greater than the diameter of the larger star. Stars this close together produce strong tidal forces that result from the gravitational pull of each of the two stars on the other. Moreover, the great speed with which the stars orbit each other causes a strong centrifugal force. The larger, cooler star may then become unstable enough that material, mostly hydrogen, will stream away from it. Eventually much of this material will form a disk of matter around the white dwarf. This "accretion disk" orbits the dwarf rapidly, and it contains a "hot spot" where new material from the cool star's stream is striking.

Fig. 18.1. SS Cygni: 21h40m.7, $+43°21'$; circle is 1°; U Geminorum type variable; range 8.2–12.1; period averages 51.6d.

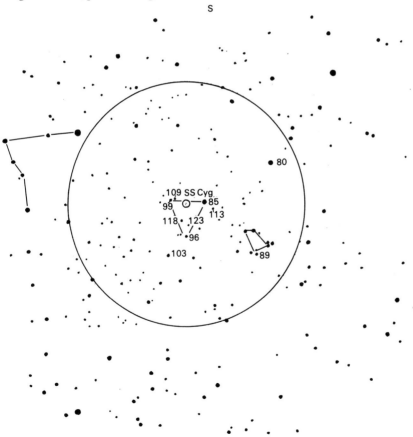

It may be that the cause of the outbursts lies in the outer part of this accretion disk. When sudden changes in the viscosity of this critical region occur, runaway thermonuclear fusion occurs and the star brightens explosively.

In SS Cygni, these outbursts occur at roughly 52 day intervals.

Fig. 18.2. Light curve of SS Cygni, based on AAVSO and earlier observations, from 1896 to 1963. Reprinted by special permission of the American Association of Variable Star Observers, through its Director, Janet Mattei.

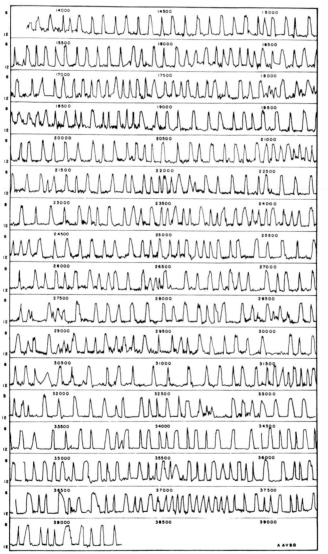

LIGHT CURVE OF SS CYGNI

However, as those of us who wager on this time know, the time of an eruption is uncertain. Also, not all eruptions are sudden; rarely it will rise more gradually.

No matter how long SS Cygni takes to go into outburst, it is still a real treat to watch it. The best way to see this most startling phenomenon is to observe the star every night. Usually, one observation is enough, but if it has been more than six weeks since the last outburst, two estimates separated by a few hours is a good idea. Steady weather patterns are vital to a successful rise-catching. From eastern Canada I observed SS Cygni for six years without once catching it on the rise. Then in Arizona, I caught it three times in a row!

But what if all you have is a 60 mm refractor whose minimum light grasp is just about magnitude 10.0, not enough to reach SS Cygni's minimum two magnitudes below? Don't give up before you try; a refractor of that size can work for this star. Remember that for all its surprises, it is predictable in one sense: over years of outbursts it has shown that if it can push itself above magnitude 11.0 it will likely go all the way.

Say last night you couldn't see SS Cygni at all. In fact, you couldn't see any star in your field fainter than magnitude 10.3. It is perfectly legitimate to record your estimate at (10.3, meaning that the variable must have been fainter than 10.3. But tonight you again set up your little telescope and a new star greets you at 10.0. Then you can be fairly certain that SS Cygni will continue to rise over a short period until it reaches its maximum of 8.3. Suppose that you are the only one to observe SS Cygni that night before it went into outburst. Your "fainter than" observation would be crucial.

More probably on one night you would not see anything and on the following night it would already be near maximum, for its rises are usually fast. Once the star is bright, you can watch it for about a week before it slowly fades to minimum.

Another note: professional astronomers, using photometers, have observed on rare occasions a rapid fluctuation of up to 0.2 magnitude while the star was near minimum. This may be caused by changes occurring in the accretion disk or its hot spot.

18.2 U Geminorum

Here is another star that truly merits watching. All of the dwarf novae are also known as U Gems, although SS Cygni has recently become so popular that it has possibly usurped U Geminorum as head of the class. In any case, while SS Cygni may occupy your summer nights, U Geminorum is surely your winter companion. Its range is 8.9 to 14.0 magnitudes. Use at least an 20 cm (8 inch) telescope and be careful of the rich star field, because it can be confusing. Remember that the 14th magnitude is pushing the visual limit of a 20 cm telescope and that your sky conditions have a lot to do with whether you will catch it at minimum. In any case, your observations don't have to include minimum sightings to be effective. Use the charts in Figs. 18.3 and 18.4 to find and observe this intriguing variable. Remember that if it appears to be going into outburst, make one estimate every hour.

Fig. 18.3. U Geminorum, AAVSO chart (b). Reprinted by special permission of the American Association of Variable Star Observers, through its Director, Janet Mattei.

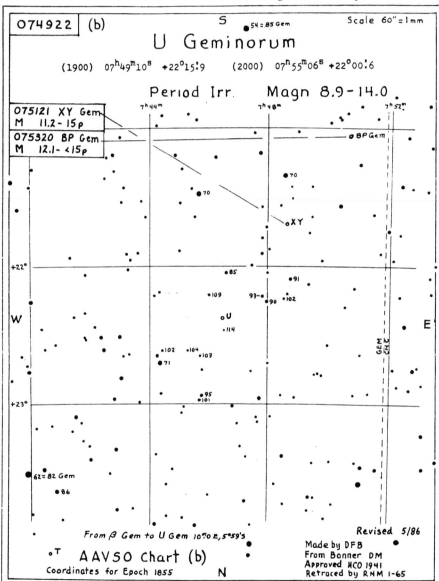

Fig. 18.4. U Geminorum, AAVSO chart (d). Reprinted by special permission of the
American Association of Variable Star Observers, through its Director, Janet Mattei.

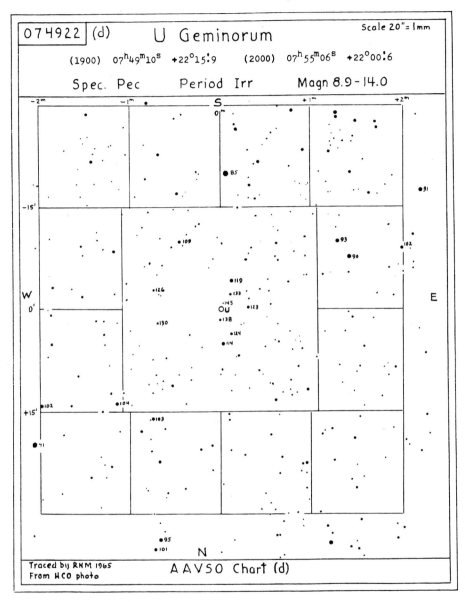

18.3 Z Camelopardalis

Like SS Cygni, Z Camelopardalis (Figs. 18.5 and 18.6) is cataclysmic, except that its range is not as great and its period is shorter. The resemblance ends there; Z Cam can shine at constant brightness, sitting comfortably at 11.7, midway between its maximum and its minimum.

We know of about 30 Z Cam type stars, and they do not have fixed

Fig. 18.5. Z Camelopardalis, AAVSO chart (b). Reprinted by special permission of the American Association of Variable Star Observers, through its Director, Janet Mattei.

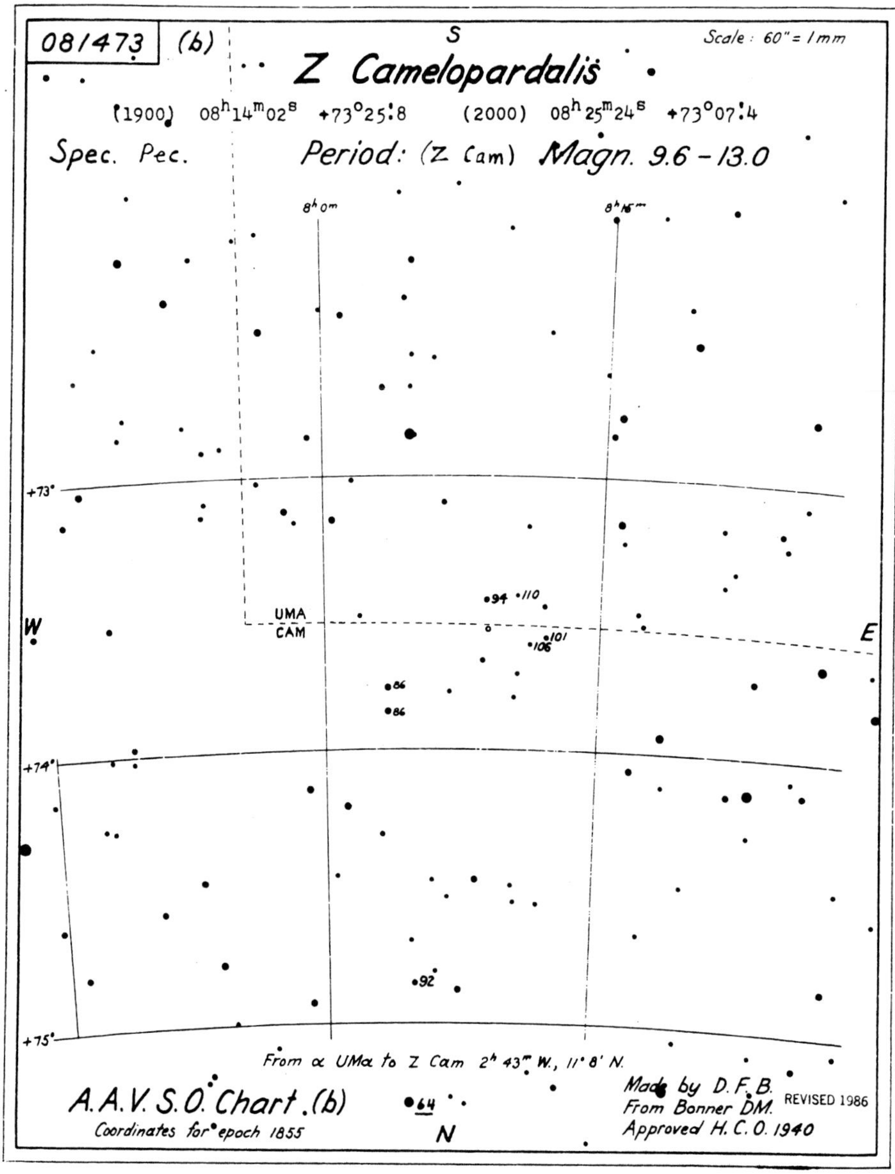

patterns of behavior. Traditionally, standstills occur when the star is on the declining end of an outburst, and when the standstill is over, the star resumes its pattern. During the late 1970s observers noticed a change in tradition with a standstill that lasted for several years. Once during this period I saw Z Camelopardalis begin to go out of its standstill and into a decline and then "change its mind" and hurry back to its same standstill position again!

Fig. 18.6. Z Camelopardalis, AAVSO chart (d). Reprinted by special permission of the American Association of Variable Star Observers, through its Director, Janet Mattei.

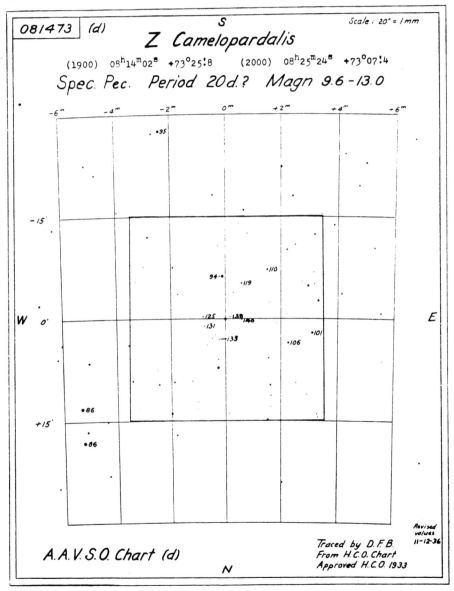

Since Z Cam's behavior is more complex than that of SS Cygni, I do recommend that you have a telescope that can see at least 12th magnitude. If your instrument is not that powerful, you might go for years without ever seeing the star. When Z Camelopardalis is in standstill, it is quite easy to estimate using the 11.0 and 11.9 standards that accompany it.

With variables, "easy" and "careless" are not synonyms. You must be careful to ensure that your estimate is as accurate as possible, for a recorded drop of 0.1 or 0.2 magnitude could indicate that Z Camelopardalis is heading out of a standstill. A careless observation might miss such a change, or worse, it could send out a false alarm. When you observe any variable star, assume that you are the only observer watching at that time, that your observation may be crucial, and that it must be made with the greatest care. This philosophy will go a long way toward increasing your cosmic credit, and months and years of carefully made observations will earn you a coveted reputation as a careful observer.

If, after trying your best to estimate Z Camelopardalis, you just cannot decide whether it is 11.6, 11.7, or 11.8, mark it as "11.7:" or "11.7?". Either mark indicates uncertainty. I have had to use this notation during hazy or even moonlit nights.

Some time ago I attended a special meeting on how valuable amateur observations on the stars have become. We had spent an enthralling afternoon listening to papers while the autumn Sun exaggerated the sharpness of the red maple leaves outside. Sitting around me were older people who had been active variable star observers for over 50 years. One of them stared in disbelief, shook his head, and said, "I had thought that the professionals had all the data they need from us. I'm an old man now. Years ago we were all part of the excitement of observing for professionals who needed our work. Then came a long period when we wondered if there was any value to what we were doing at all. Now, the old excitement, the freshness, is back again."

19   A nova in reverse?

19.1   R Coronae Borealis

Have you ever had a backwards day, when everything seemed to be happening in reverse, and things turned out to be precisely the opposite of what you expected? Such a day could only end with your first observation of a "backwards nova," a star called R Coronae Borealis.

Usually, a nova stays at minimum until the day of its mighty explosion. R Coronae Borealis does the opposite. It stays at maximum, a bright beacon around magnitude 6, and then without warning plunges eight full magnitudes to the depths of a magnitude 14 minimum. At its brightest it can be seen with a good pair of unaided eyes if the sky is dark enough. At minimum it will test the mettle of a 20 cm (8 inch) telescope.

R Coronae Borealis is not really a nova in reverse; it only acts like one, perhaps by surrounding itself at completely irregular intervals with a shell of carbon particles which absorb light. In late February of 1977 I returned from an evening out and the clouded sky was just beginning to clear. I decided to check the sky and observe just one variable for a total session of no more than ten minutes. I thought I would try R Coronae Borealis since the night before and for many months earlier it had been shining brightly at about magnitude 6. It often happens that my shortest observing sessions turn out to be the most productive. That night R Coronae Borealis

Fig. 19.1. R Coronae Borealis: 15h46m.5, $+28°18'$; circle is 5°; R Coronae Borealis type variable; range 5.8–14.8.

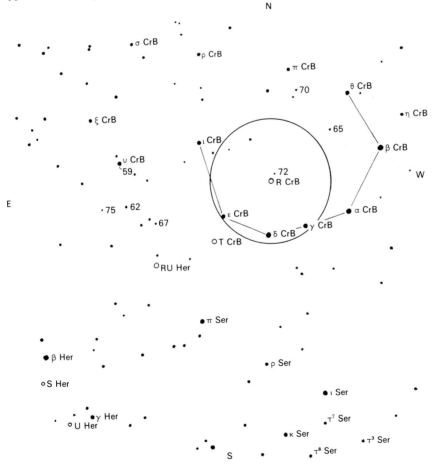

had slipped to magnitude 6.4, ready to plunge to a minimum of 14.3 that would last the rest of the year.

The mystery of R Coronae Borealis is surpassed only by the beauty of this star and its surroundings (Figs. 19.1, 19.2, and 19.3). Located right in the middle of that semicircle of stars called the Northern Crown, the star at maximum shines at about 6th magnitude, just barely visible with a sharp unaided eye during a dark night. Through binoculars, R CrB is near

Fig. 19.2. R Coronae Borealis: circle is 2°.

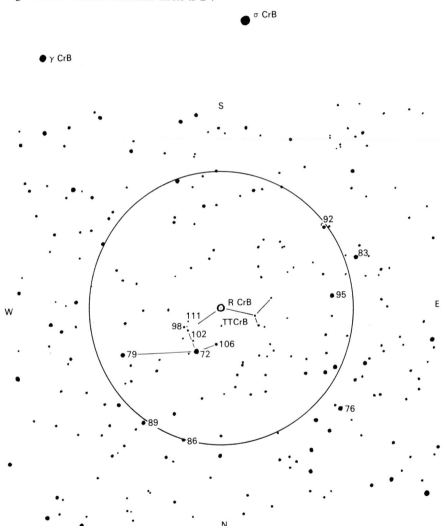

two fainter stars, one a 7.2 magnitude sun and the other shining at 7.8. I have used the 7.2 star as a test for pristine dark sky conditions; occasionally on dark Arizona nights I can see it without optical aid.

To take R CrB seriously means to observe it every clear night. The start of a minimum is slow and teasing; R CrB can drop as low as 6.4 without necessarily plunging all the way to minimum. But if your first observation is 6.2 or 6.3, and the following night you find a 7.1 star, then the chances are good that the star has begun a decline that will not end until you need a 30 cm (12 inch) telescope to see it.

19.2 RY Sagittarii

Like R Coronae Borealis, 191033 RY Sagittarii spends much of its time at its bright 6.5 maximum. At the end of the 19th century, a drop in brightness was noticed by Col. E. E. Markwick from Gibraltar, and

Fig. 19.3. R Coronae Borealis: circle is 0.5°.

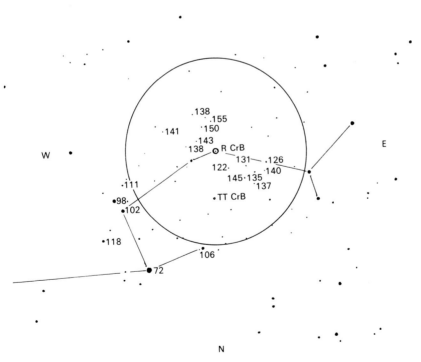

amateur observers have kept this star under close watch ever since. Fig.
19.4 is a history of the antics of this strange star from 1892 until 1972. A
look at this light curve will show that the drops are unpredictable both in
their frequency and in their depth; although the minima can go as deep as
14.0, quite often RY Sgr does not drop by nearly that amount.

Fig. 19.4. Light curve of RY Sagittarii, an R Coronae Borealis star, based on AAVSO
and earlier observations, from 1892 to 1972. Reprinted by special permission of the
American Association of Variable Star Observers, through its Director, Janet Mattei.

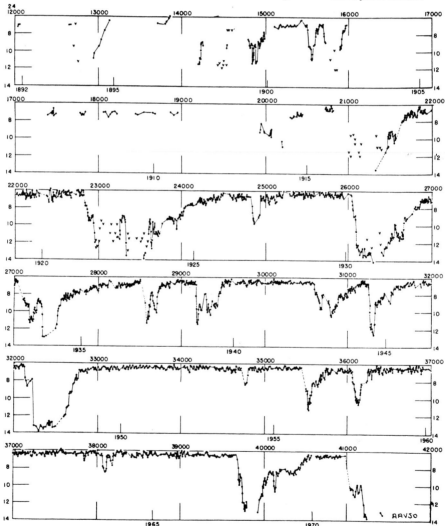

20   RU Lupi?

Probably.

Ask your friend. Especially if by now you are an avid variable star observer, the answer will be yes! So let's take advantage of the situation and perambulate amongst two of the most unusual objects in our corner of the universe, 041619 T Tauri, and 155037 RU Lupi, two stars that represent the carefree conduct of cosmic youth. Both stars are irregular variables and represent stars that are observed in clouds of gas called diffuse nebulae.

20.1   T Tauri

Let's begin with T Tauri, the easier of the two, a star that can be watched easily from northern latitudes in the frigid evenings of winter.

If you can find the Hyades, you can find T Tauri. Moving from Delta to 64 Tauri to 4th magnitude 68 Tauri, you proceed to the northeast until you reach Epsilon Tauri, a 3.6 magnitude star (Fig. 20.1). Slightly to the southwest of Epsilon is a group of four stars in the shape of a malformed kite. I use this group as a guide to T Tauri.

Notice the magnitude range of 9.3 to 13.5; I have observed T Tauri for several years and I have never once seen it go fainter than magnitude 10.6 nor brighter than the mid 9's. Most observers have reported it hovering just brighter than 10th magnitude.

But that's only part of the story. T Tauri occasionally can flicker. On the night of November 9, 1978, I noticed T Tauri jumping wildly between magnitude 10.0 and 10.5, varying by as much as 0.2 magnitude in less than a minute. Could it have been haze, light cloud cover, or tired eyes? I tested for these by running checks on the nearby comparison stars. They were shining peacefully at their assigned magnitudes. Only T Tauri was performing. On only one other occasion did T Tauri show signs of such rapid flickering, but never so much that it matched that cold November night, when I had to estimate every 60 seconds just to keep up with the moods of this cosmic candle.

There are two ways to estimate T Tauri's magnitude. For long term variations, you should make an estimate every week or so. But because T Tauri can flicker, the AAVSO recommends that you make a train of four or five estimates per night, each separated by 10 or 15 minutes. You will probably find that the magnitude 9.3, 9.8, and 10.8 comparison stars are the most helpful. As long as the star is in the same range of half a magnitude, try to use the same comparison stars, for if you change your set of standards without good reason you could change the meaning of your estimates.

20.2 RU Lupi

Now we are ready for RU Lupi, a star with the same basic behavioral patterns as T Tauri. The chart for RU Lupi contains no comparison stars; you have only a means to find the star. To observe it you will be a pioneer and create your own comparison sequence.

Fig. 20.1. T Tauri, AAVSO chart (b). Reprinted by special permission of the American Association of Variable Star Observers, through its Director, Janet Mattei.

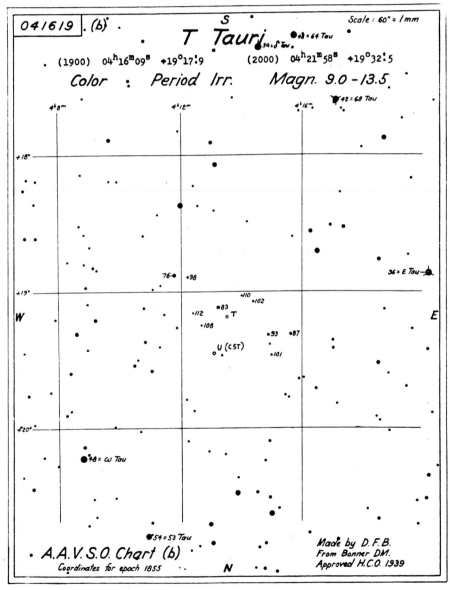

Unless you live south of 40 degrees north latitude, you will have problems even finding RU Lupi. Once you have located it, make your own drawing of the surrounding area, and use the fractional method, explained in Chapter 8, for estimating. Choose five stars near in brightness to the variable and letter them *a* through *e*, brightest to faintest. When you estimate, judge the relative brightness of RU to any two consecutive stars in your sequence. If you decide that the brightness if RU is midway between stars *a* and *b* then write down "*a*,1,*V*,1,*b*." If it is three-quarters fainter than star *a* and a quarter brighter than *b*, you record "*a*,3,*V*,1,*b*."

Designed to give you an idea of how to watch an infrequently observed star for which comparison stars have not been "officially" chosen, this RU Lupi project is a challenge. Since this star's variation, like that of T Tauri, is irregular, the actual star magnitudes are not as important as understanding its behavior relative to the comparison stars you have selected.

Fig. 20.2. Finder chart for 155037 RU Lupi, marked as +. T Tauri type variable. Chart covers 4 × 4 degrees; north is up, east to the left. The bright star to the southeast of RU is Eta Lupi, magnitude 3.4, and the bright star near the northeast corner is Theta Lupi, magnitude 4.2.

21   Orion, the star factory

Buried deep in the richest part of the winter sky is the magnificent constellation of Orion the Hunter, a group of stars that have a closer relationship than the chance positioning of most constellations. Except for Betelgeuse, most of the bright stars in Orion are roughly the same distance from us. Orion is far more than just a place in the sky, named for a figure from mythology. Orion is a cosmic production plant, whose different divisions show us how stars are made.

The process of star formation is very complicated, but here we can see how it takes place. Young stars in the belt of Orion represent a group of stellar children whose formation is essentially complete. The sword area is a cosmic nursery with some of the stars being less than a million years old. Some astronomers suspect that the area south of the sword, with its hydrogen-rich regions, given another several million years, will coalesce to produce new stars.

In the area where stars are in the process of birth, the highlight is M42, the Orion nebula. The darker your sky, the greater the thrill of such a sight. Here is a nebula whose visual appearance is stunning, whose delicate patterns are exquisite. With a 10 cm (4 inch) telescope, you can make out the beginnings of the complex light and dark gases that share the nebula. With a 20 cm (8 inch) telescope and a dark sky, the region explodes into a labyrinth of convoluted gases that surround a hundred suns. If you can get to the eyepiece of a 40 cm (16 inch) telescope, you will find how the delicacy of these strands of nebulosity is overwhelming.

Color is easily seen in a telescope of such aperture, the northeast area especially, assuming a faint but noticeable fuchsia. It is hard to look at M42 under ideal conditions and not be profoundly moved by the splendor of a cloud of gas 25 light years across, shining from a celestial stage 1600 light years away from us.

The young variable stars are members of what we call a "T association" of stars that have formed from a common cloud of dust and gas. They are an intricately woven portion of the cosmic fabric of the Orion nebula. The Orion stars vary for several reasons, one of which may be their passage through differing thicknesses of nebulosity. These stars are also believed to vary intrinsically. The result is a group of stars whose complicated patterns of behavior display two or more types of variation.

These nebular variables are not, I must caution, traditional stars with long, well established periods. They are young stars that flicker by a half magnitude or so over a period of a few minutes. Because they are embedded in the uneven background of the Orion nebula, they are very hard to estimate. Try every 10 or 15 minutes, being careful not to be fooled by the cacophony of other stars in the nebula which may be varying also. Always use, for each variable, the same comparison stars, and estimate only on nights when no cirrus clouds are present.

These are not simple stars to observe, but together they perform as a symphony of delicacy and brilliance. From 1978 to 1981, observing the Orion variables was the project on which I spent most of my observing time. To get a good understanding of how these stars vary, I made over 10,000 observations, mostly through an 8 inch (20 cm) reflector. I felt I was visiting a family at its dinner table, watching its changes and being a part of its moods.

21.1 V351 Orionis

V351 Orionis is an unobtrusive, innocent looking star located very close to the fan-shaped nebula M78 (Figs. 21.1 and 21.2). It should be watched every 10 or 15 minutes over a one or two hour period each night. It is one of the infamous sirens of Orion. This fickle star beckons with the promise of lots of action in a short time; however, if you are not careful, the flickers and other brightness changes may be imaginary, and there is no way of checking with predictions. Unfortunately, there is no "trick" or specified observing technique short of comparing your work with that of other observers to ensure accuracy.

Less than a quarter degree from V351 is GT Orionis, a semiregular variable that you may enjoy observing every two weeks or so. The magnitude range suggested on the charts was derived photographically (hence the *p* after the magnitude range); your visual range will be a bit brighter.

I suggest that if you plan to include V351 in your program, get as much experience observing it as possible *before* you begin to file your reports. After two or three months of nightly checks, you should have an idea of what constitutes variation by about a fifth of a magnitude. As with eclipsers and very red stars, watch judiciously for the influence of moonlight, cirrus clouds, smog, and those deadly street lamps, all of which add to a witches' cauldron of perils.

V351 Orionis, by the way, happens to be one of the more straightforward of "nebular" variables, as the Orion, or "T Tauri" variables are often known. The Orion nebula itself contains hundreds of similar young stars, some of which I have been observing for years. But some of the M42 variables are deeply involved in dense nebulosity. They are so difficult to observe and are subject to so many observational errors that I wouldn't recommend them to any but the most seasoned and patient observers.

These dire warnings are not intended to turn you away from variables but to reinforce your understanding that this field is worthy of careful attention and accurate observation.

21.2 How Orion stars vary

The most common type of Orion star behavior is represented by a rapid and highly irregular flicker. The stars can vary by as much as 0.2

magnitude in 20 minutes. For this reason, it is necessary to watch these stars very closely. I recommend making one estimate every 10 or 15 minutes, and that you choose comparison stars with great care, since some of them may themselves be variable.

The second pattern of variation reminds me of R Coronae Borealis, in which long periods of maximum light are interrupted by unpredicted sudden drops to a much fainter magnitude. T Orionis behaves this way quite frequently and at one time was even classified as an R Coronae

Fig. 21.1. V351 Orionis, AAVSO chart (b). Reprinted by special permission of the American Association of Variable Star Observers, through its Director, Janet Mattei.

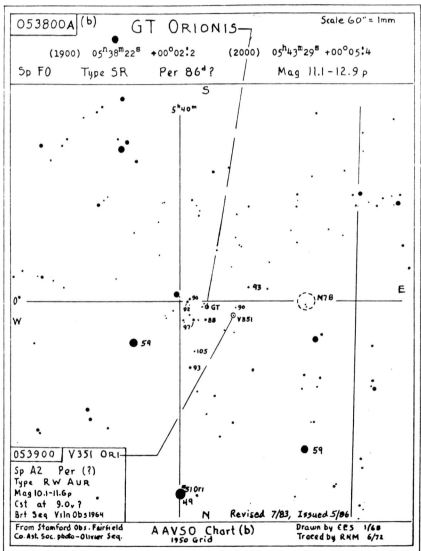

Borealis star. But just like two ailing people whose similar symptoms can be caused by different illnesses, the variation of R Coronae Borealis and T Orionis have different causes. The large drops in the brightness of T are possibly caused by thicker amounts of nebulosity passing in front.

Other Orion stars have much flatter light curves. In fact, MX Orionis has a "light straight" — the five years I have observed it, MX has not

Fig. 21.2. V351 Orionis, AAVSO chart (d). Reprinted by special permission of the American Association of Variable Star Observers, through its Director, Janet Mattei.

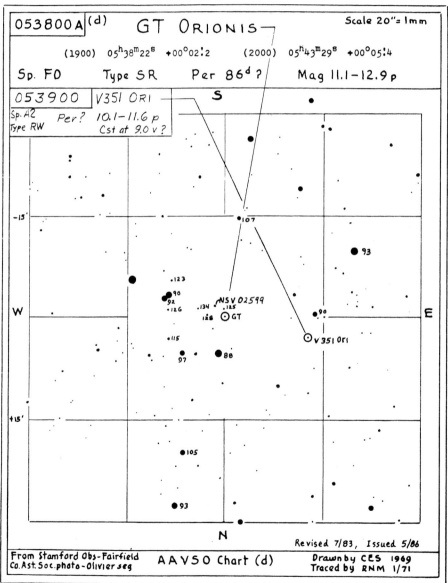

varied by more than 0.2 magnitude. Also, there is V361 Orionis, which is embedded in such bright nebulosity that it is really impossible to estimate its brightness visually. The nebulosity gets in the way, confusing both the mind and eye.

Estimating these stars is a very difficult process, so exacting for the limited resources of the human eye that the AAVSO no longer encourages observation of these stars except by their most experienced observers. So why bother with them at all? We bother because Orion represents an important phase in the early evolution of a star. You will be watching these stars not for any variable star association, but for yourself. Even though your results go no further than your eye and notebook, your work will be an attempt to understand how young stars behave. Remember your purpose: to determine small differences in magnitude, so that an observed *change* of 0.2 magnitude is more significant to record than the actual magnitude numbers themselves. Whether NV Orionis is at 10.1 or 9.9 is not as important as your correct perception of a change in brightness. Be consistent; always use the same set of comparison stars. Concentrate on the stars you think are active. Finally, don't forget to enjoy the enthralling nebula of which they are a part.

21.3 Other Orion variables

Here are other Orion variables worth watching. They all share designation 053005. Use the charts in Figs. 21.3 and 21.4 to find them. Fig. 21.5 summarizes information we have on their ranges of variation.

HU Orionis. Use a 30 cm (12 inch) telescope. Star shows little activity, usually not more than 0.5 magnitude.

IY Orionis. With a range of 12.1 to 13.0, a 30 cm telescope is recommended.

V372 Orionis. Range 7.8 to 8.5, so a 15 cm (6 inch) telescope is fine. Since the comparison stars are distant, do not use an eyepiece that gives too narrow a field. Variations are irregular, and not strong.

KS Orionis. Easily found with an 20 cm (8 inch) telescope, but shows little variation.

LP Orionis. This star has shown rapid and strong variations, but it is embedded in such dense nebulosity that it is hard to observe.

MR Orionis. Faint, so use at least a 30 cm telescope. This star shows considerable variation, but it is close to the center of the nebula, so use the same power and the same telescope to avoid different levels of contamination by the nebula that would result from the use of different telescope fields.

MX Orionis. This star shows practically no variation at all from visual observation.

NU Orionis. Although it is officially listed as having possible strong variation, I have not observed this during six seasons of viewing. It is bright, so use a small instrument.

V361 Orionis. Possibly the most interesting of the brighter variables in M42, this star shows occasional strong variations. The problem is that it is too close to the center of the nebula for reliable visual observation, unless you are careful to keep the conditions of telescope, eyepiece, weather, and moonlight as constant as possible. Use at least a 15 cm telescope.

NV Orionis. This star has shown significant and rapid change, by as

Fig. 21.3. Nebular variables in the region of Theta Orionis, AAVSO chart (d). Reprinted by special permission of the American Association of Variable Star Observers, through its Director, Janet Mattei.

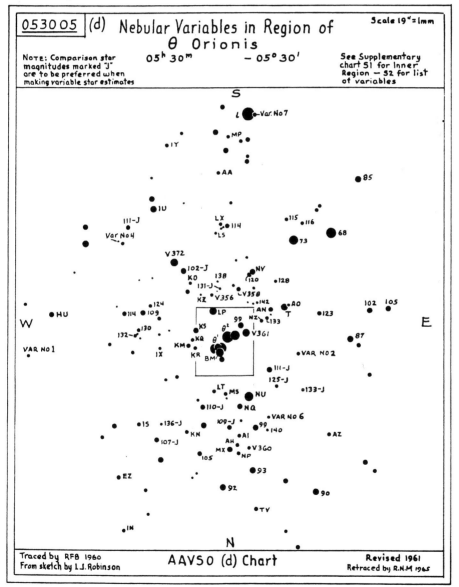

much as a fifth of a magnitude in 20 minutes. On the night of September 14/15, 1977, I watched this star brighten from 10.1 to 9.7 over four observations spanning 30 minutes. Use at least a 20 cm telescope.

No. 2 Orionis. This bright 9th magnitude star is not formally named because its variation has not been confirmed. It may vary by a few tenths of a magnitude. Use a 15 cm (6 inch) telescope.

Fig. 21.4. Nebular variables in the region of Theta Orionis, AAVSO chart (S1). Reprinted by special permission of the American Association of Variable Star Observers, through its Director, Janet Mattei.

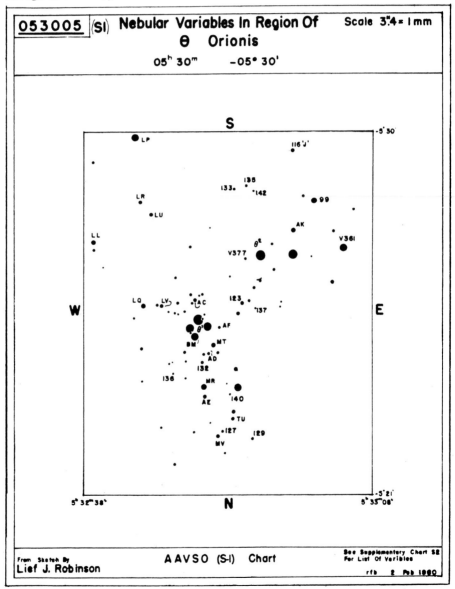

21.4 Eclipsing binaries in M42

Theta Orionis A. The westernmost member of the trapezium of Orion is such a variable, discovered by Eckmar Lohsen. Marvin E. Baldwin of the AAVSO has calculated a period of 65.432 days. Its range is 6.7 to

Fig. 21.5. List of nebular variables in the region of Theta Orionis, AAVSO chart (S2). Reprinted by special permission of the American Association of Variable Star Observers, through its Director, Janet Mattei.

053005 S2						NEBULAR VARIABLES IN REGION OF					

θ ORIONIS

05ʰ 30ᵐ   -05° 30'

Var.	R.A. (1950) Dec.			Max.	Min.	Var.	R.A. (1950) Dec.			Max.	Min.
	h m	s	o '				h m	s	o '		
HU	5 31	11	-5 29.1	11.3	12.1	AF	5 32	52	-5 25.2	12.2	(14
EZ		51	06.9	11.2	12.6	TU		53	22.9	11.6	(14.5
IN		54	-4 59.8	12.9	13.3	MX		54	11.2	9.6	10.5
IS	32	03	-5 13.8	12.6	(15.5	V377		55	26.9	12.8	14
IU		09	43.7	8.8	10.0	V358		57	32.7	11.9	12.4
IX		14	24.7	12.6	13.8	AH		58	11.8	12.2	13.9
IY		15	52.3	12.1	13.0	NP		59	10.6	11.5	12.6
V372		21	36.3	7.8	8.5	NQ		59	17.1	11.1	12.4
KM		29	25.2	11.3	12.2	AI	33	00	13.1	12.0	14.2
KN		30	13.6	12.8	13.7	AK		00	27.6	11.3	14.0
KO		30	33.6	13.0	14.0	NU		04	18.0	6.5	7.3
KQ		31	25.9	12.3	13.5	V360		05	11.4	12.5	13.5
KR		33	25.0	12.3	14	V361		05	27.2	7.8	9.6
KS		33	27.3	9.9	10.9	NV		05	35.1	9.5	11.3
KZ		37	32.0	13.0	13.7	TV		09	03.2	12.9	15.1
LL		39	27.3	10.9	12.6	NZ		11	28.5	11.9	14.2
LP		43	29.9	8.0	9.4	AN		15	30.1	10.5	12.1
V356		43	32.0	12.8	13.8	T		24	30.5	9.4	12.2
LQ		44	25.7	11.8	13.0	AQ		25	30.7	13.	14.2
LR		44	28.3	11.9	13.2	AZ	5 33	50	-5 13.5	12.2	13.1
LS		44	40.4	12.6	13.3						
AA		44	48.5	12.3	13.7						
LT		45	18.9	12.7	13.5	Suspected Variables					
LU		45	28.0	12.0	13.4		h	m	o '		
LV		46	25.7	12.1	13.3	Var #1	5	31.0	-5 23		
						#2		33.5	-5 24		
LX		47	41.5	11.9	13.1	#3		32.8	-5 25		
BM		49	25.1	8.1	8.8	#4		31.7	-5 38		
AC		49	25.8	12.5	15	#6		33.2	-5 16		
MP		49	53.7	12.5	13.5	#7		33.0	-5 56		
AE		50	23.5	12.3	(14.1						
MR		50	23.7	10.3	12.0						
MS		51	18.6	12.8	13.5						
AD		51	24.5	12.9	15						
MT		51	24.7	11.2	13.0						
MV	5 32	52	-5 22.5	11.7	13.2						

AAVSO   (S2)   CHART   April 1965

7.7. The period of eclipse lasts some 20 hours, which includes 2.5 hours of minimum light.

BM Orionis. In the Orion nebula, but apparently not sharing the characteristics of the other nebular variables there, BM Orionis is a traditional eclipsing binary with a period of 6.5 days. The drop in brightness is just slightly over half a magnitude, from 8 to 8.6. The eclipses are longer than those of most of the other stars we have examined, about a day. Moreover, the time of deepest eclipse lasts nine hours. In 1917 Fritz Goos discovered that this star varies, and Ernst Hartwig later proved its eclipsing nature. Apparently no observer has reported nebular variations.

22   Other variable things

What else does the sky have that varies? Besides stars, there are other objects worth noting. Some are easily visible, and others are beyond the reach of all but the most powerful telescopes and detectors. They range in distance from a few million kilometres to billions of light years.

22.1 Variable nebulae

In the constellation of Monoceros is a variable star called R. It varies irregularly by about half a magnitude around 11. However, the star is usually very hard to see. The reason is that it is embedded in a nebula which also varies in brightness! This object is known as Hubble's variable nebula, NGC 2261, after the Mount Wilson astronomer who in 1916 discovered that it varies in brightness, size, and even shape. The variation does not seem to follow the brightness changes in R Mon, and they do not occur with any regularity.

R Monocerotis and its nebula probably represent a planetary system in an early stage of formation. At least two other variable nebulae are known, NGC 1555 in Taurus, and a tiny wisp in Corona Austrina, NGC 6729, the home of R Coronae Austrinae. (See p. 151.)

22.2 Active galaxies

Innocently displaying some irregular brightness changes are a number of objects that have recently been identified as the cores of galaxies. The Seyfert galaxies are spiral galaxies with starlike nuclei that are very bright and slightly variable. They vary in leisurely fashion with ranges of up to three magnitudes over periods of months or years, although much faster changes of a fraction of a magnitude have been observed over a period of days. Messier 77 is an example of this class of galaxy.

When Carl K. Seyfert discovered this type of galaxy in 1943, neither he nor anyone else could foresee its future importance. In June of 1960 he died in an auto accident in Tennessee. The obituary in *Sky and Telescope* from that August summarized his rich career without mention of the Seyfert galaxies.

A quasar, short for quasi-stellar object, is believed to be the highly energetic core of a *very* distant spiral galaxy. If the galaxy were "normal," like M31, we would expect to see nothing because of its great distance. But the active part is visible as a faint starlike object. Unlike the nuclei of Seyfert galaxies, those of quasars can be thousands of times as bright as the surrounding galaxy. The nature of the quasars was first suggested in 1963, and is still under debate.

Observationally, quasars can be treated like semiregular variable stars. Their variations are slow and irregular, but they apparently change brightness just like slowly varying stars.

The brightest known quasar, 3C-273 (object 273 in the *Third Combined Catalog of Radio Sources*), appears as a starlike object of about 13th magnitude. I have seen it with a 15 cm (6 inch) reflector, finding it just as I would find an ordinary variable. I then began to make the estimate, as I would estimate an ordinary variable, when I realized that the little point of light may be the most distant object I have ever seen, the core of a galaxy several billion light years away. With a sobered and respectful feeling, I continued the observation.

Estimate this quasar once or twice a month. Although its variation is slight and visual observations are of doubtful value, it is fun to look at an object that far away.

There are also some elliptical galaxies with very bright, starlike nuclei. Since the active nucleus was all that was at first visible, these were treated as irregular variable stars until their great distances were recently revealed. These BL Lacertae objects vary dramatically by as much as two full magnitudes in a single day! They are named for the first discovered, which was initially mistaken for a variable star.

22.3 Minor planets

We suddenly come close to home, to some small "primitive bodies" that orbit the Sun, mostly between Mars and Jupiter, although some approach to within a few million kilometres of Earth and a few inhabit the realm of the giant planets.

Watching minor planets, also called asteroids, can be a fascinating pastime. You can begin with the four largest, and first discovered, bodies: Vesta, Ceres, Juno, and Pallas. These are easily visible not just during Earth seasons, like variable stars, but when their orbital "seasons" are favorable too. Unless they are near "opposition" to the Sun's position in the sky, and moving westward, they travel in an easterly direction across the sky. Thus their seasons are long, but not annual.

Occasionally a minor planet will traverse the field of a star cluster, or
will pass near a bright star. If such a circumstance is about to happen the
astronomical magazines usually inform their readers, providing conveni-
ant finder charts. That is a good time to get acquainted with a new minor
planet. When asteroid No. 15, Eunomia, passed in 1977 before the
Beehive Star Cluster in Cancer, I spent many nights watching it creep past
the member stars.

Minor planets vary in brightness. As they rotate, they present different
aspects of their elongated bodies towards Earth so that we see brightenings and fadings of up to a magnitude. We can estimate asteroids precisely
the same way we estimate variable stars, by comparing their brightnesses
to those of comparison stars of known magnitude. The only difference is
that the asteroid does not stay in its field for more than a short time, a
matter of days or weeks. Also, the range of variation in most cases is a
portion of a magnitude, which means that visual observations are
difficult. Frequently a minor planet will cross a star field that the AAVSO
has charted. If you have a copy of the *AAVSO Variable Star Atlas* you will
finds its charts filled with comparison stars.

If you succeed in getting a good comparison sequence, estimate the
object every half hour or so. If your observing session produced little
evidence of change, the asteroid's range may be slight or its period may be
long. Eunomia's rotation is just over six hours, which includes two
maxima and two minima, but others take as much as several days to
complete a rotation. If the aspect of the asteroid's orbit is such that we are
looking along its axis of rotation, we may not detect any evidence of
brightness change no matter how fast it rotates.

22.4 Comets

The mysterious, majestic comets offer so much for a visual
observer that I am sorely tempted to depart from the single subject of
variation. Even the simple estimating of the magnitude of a comet is a
fascinating experience.

Comets vary for many reasons. Their approach to Earth will result in an
apparent increase in brightness, and approaching the Sun will usually
result in a real increase in brightness as the comet is physically affected by
the Sun's heat. A comet does rotate, but determining that rotation
visually is difficult since we do not see the nucleus from our groundbased
telescopes. However, when a bright comet is near the Sun it often sends
out jets of dusty material that change radically from night to night,
resulting in changes of brightness of up to a magnitude. If you can
demonstrate that these jets are recurring at regular intervals, that might
provide a hint of a rotation period.

Do not estimate a comet by comparing its brightness to galaxies or
other non-stellar objects. Comets are not galaxies. To estimate a comet's
brightness, defocus your telescope until the comet and comparison stars

you have selected have a similar apparent size. Estimate these out-of-focus "balls of light" by going back and forth between the brighter and fainter comparison stars, and comparing each with the in-focus comet. Your estimate improves if you try this with at least one other pair of stars.

23   The Sun

Variation on the Sun! For many years the AAVSO has had a section for observation of the Sun, the logic being the star around which we revolve is a variable. In a stretched sense this may be true, but if we were to observe the Sun as we watch other stars, from light years away, we would find it shining at a constant brightness, without any indication whatsoever of its 11 year cycle of variation that we see manifested in sunspot activity.

Whether we worship it, plan our lives by its schedule, tan ourselves by its light, bask in its warmth, or study it, the Sun is a star whose importance cannot be overstated. And when we observe it through our telescope, we learn much about the the churning, changing nature of the star around which our planet turns.

An amateur astronomer and pharmacist of Dassau, Germany, Heinrich Schwabe, discovered the Sun's "variation" in 1843 through his long series of meticulous observations of its activity. After buying a small telescope he began to search for a planet inside Mercury's orbit, hoping to find it transiting the Sun's surface. This "Vulcan" idea still lives, and as late as 1982 infrared searches have attempted to find such a planet. It has not been found. Schwabe's serendipitous discovery was the 11 year sunspot cycle.

The most obvious solar feature is the sunspots, magnetic storms on the solar surface that appear dark because they are cooler than the rest of the surface. These spots appear in groups, and if you look carefully around these groups, you may notice some brighter regions called faculae.

By observing the Sun every day, you can repeat this epochal discovery for yourself, and you do not need a large telescope. In fact the small refractor that has been gathering dust in your attic could be just the thing for this work. Its long focus will provide a field just wide enough to show the entire Sun.

Solar observers cannot take too many precautions for their safety, as a fraction of a second look at the Sun through a telescope would cause permanent blindness. You can avoid this disaster by using a safe way to find the Sun, and one of two safe ways to look at it.

How do you safely find the Sun? First, do not use the finder. Make sure, in fact, that the telescope's finder is either covered or removed so that no one accidentally looks through it. Find the Sun by moving the telescope until its shadow is as small and directed as possible. The Sun should then be close to shining its rays right down the tube.

23.1 Observing the Sun

Projection

Let the Sun's rays travel directly through the objective, out of the eyepiece, and into a box with white cardboard at one end. The cardboard serves as a screen for the performer Sun, and the box, properly oriented, provides shade and contrast for the image. Use an inexpensive eyepiece since the heat concentrated by the objective can damage it.

An advantage of this procedure is that a large group of people can see the Sun at one time, thus enabling them to repeat the exercise each day to observe change.

Objective filter

I recommend a filter designed to fit over the front end of your telescope, thus blocking most of the Sun's rays before they undergo any focusing at all. This produces a safe, comfortable view. Do not even think of using the eyepiece filter that comes with many small refractors you buy at department stores. These filters work at the worst possible place, attaching right onto the eyepiece and trying to reduce the Sun's light where it is most strongly focused and hottest. Chances for breaking the filter are greatest there, and a single broken filter can ruin your eye permanently.

You can purchase safe objective filters from several commercial sources. They use a coating of metal either on glass or plastic. Do not use any of the standard photographic filters, as they are not strong enough.

23.2 Projects

Daily sunspot count

Sunspots are arranged in groups and in most cases it is not difficult to determine where one group ends and another begins. Normally characterized by at least two large spots, known as a "preceding" and a "following", sunspot groups often have a cacophony of smaller tagalongs. Each day, record the number of groups, the number of spots in all of the groups.

The AAVSO has a special section devoted to solar observing. In their forms, they ask for "R numbers" for the Sun each day. The R, or relative sunspot number, is calculated simply by adding ten times the number of groups to the number of spots. It sounds complicated, but it simply is a statistical way of determining the solar spot activity for a given day. It has been done since the early days of sunspot counting.

The Sun's morphology

The Sun is presently the only star whose features we can study in detail. On a sheet of paper, draw a 6 inch (15 cm) diameter circle, and inside the circle draw the groups and spots as you see them through your protected telescope. You may also see bright spots, or faculae that you would want to draw as well. When your drawing is completed, determine the directions of east and west by moving your telescope back and forth slightly on its right ascension axis. You do not need an equatorial mount for this; the alternative is to let the Sun tell you its cardinal points itself by drifting through your field of view. The motion of a sunspot will then give away the east to west line, and then you know how the sunspots are oriented.

In greater detail, you also can draw a single group or a large spot, and code the spot you choose to draw to its corresponding appearance on the full disk drawing.

The aim of drawing is not to make an artist out of you; instead, your purpose is to record data as accurately as possible, and through such careful attention, to increase your understanding of the changing face of the Sun.

Using an objective filter and Kodak 2415 Technical Pan Film, try to photograph details on the Sun's surface. Try for the shortest exposure possible; with the Sun there should be no shortage of light!

Although they are expensive, filters that allow you to see the Sun in "hydrogen alpha" light are available commercially. These filters allow you to observe features on the Sun that are not visible in "white" light. You can see the prominences that erupt on the visible edge, or the filaments, as prominences are known when we do not see them at the edge. These eruptions are dynamic, difficult to predict, and always different. Major brightenings on the Sun's surface, known as flares, can also be seen this way. Rarely a flare will be so prominent that it can be seen without such a filter.

23.3 The Sun's future

Now that we understand that the sky is full of stars that vary, we would certainly not wish to take our own Sun for granted. For most of its almost five billion years it has given steady light to its orbiting family of major and minor planets, moons, comets, and meteoroids. Even though observations from the ground and from satellites in space have indicated a barely noticeable drop in the Sun's output during the first half of the 1980s, the Sun's basic integrity is sure to last for at least 5 billion more years, as long as the supply of hydrogen fuel lasts.

The first changes will occur as the hydrogen supply becomes seriously depleted and the core is filled with helium, the product of hydrogen fusion. As fusion burning slows down, the core cools, allowing gravity to take over and contraction to begin. As the core gets smaller, its temperature rises, and eventually it becomes so hot that some hydrogen surrounding it in a shell is ignited, and begins to expand. As the hydrogen shell gets larger, it cools, becoming red. Our Sun has swelled into a red giant, has lost the stability of youth, and could possibly start to vary as a Mira star that some extraterrestrial amateur astronomer could add to monthly observing totals. Surely we could not do that, for as the Sun grew in size, its heat would scorch our planet, extinguishing its delicate life. As the Sun continued to expand, it could even engulf the Earth.

After some periodic contraction and expansion, the Sun's outer layers would gradually escape into space, exposing the Sun's dense and slowly cooling core that would continue to shine as a white dwarf. For a short time, temperatures on Earth may even return to seasonal normals, so to speak, but that would not last long. With nuclear fusion over, the Sun would simply get cooler until its light flickers and is extinguished, ending the story of our solar system.

3

Suggested variables for observation throughout the year

24 Introduction

This part of the book contains stars that have not been described in earlier chapters. It is intended to help you plan an observing program by introducing you to a selection of interesting variable stars. By reading through these chapters, you should find some stars that you will enjoy watching. The finder charts are intended to help in finding the location of a variable star. Once you have decided on a program, I suggest that you order a complete set of charts for each star you choose and be careful to plan in advance the best time and equipment for observing them. The order in which the different constellations are presented in each chapter represents a vague and somewhat arbitrary eastward movement across the sky.

For each star, I have included the range, period and a code that specifies level of difficulty:

    1 = very easily found and estimated
    2 = a good star for beginners
    3 = some challenge, either in finding or in estimating
    4 = quite difficult
    5 = recommended only for advanced observers with larger instru-
        ments

Different sources provide different values for maxima, minima, and ranges of many variable stars, especially those with uncertain variations. In most cases I have used the values given in the *General Catalogue of Variable Stars* by B. V. Kukarkin *et al.*, Fourth Edition. For Mira stars, I have used the typical ("mean") maximum and minimum values published in the Remarks section of the Third Edition of the *Catalogue*, rather than the extreme values listed in the tables. (The means do not appear in the Fourth Edition.)

Even though finder charts are provided, you will find a star atlas extremely helpful. Tirion's *Sky Atlas 2000.0*, *Norton's Star Atlas*, and Dickinson's *Edmund Mag Six Star Atlas* are all described in the bibliography.

The scale of each chart is shown by a five degree line at the bottom.

25   January, February, March

In cold weather, we draw inward as the frigid weather and short days beckon us away from the stars and towards the armchair. This is an unfortunate loss for northern hemisphere observers who forego the unparalleled richness and diversity of the sky at this time of year, a sky that dares us to defy the inside comforts and go outside and watch. This is a time of challenge.

With its stunning belt and sword regions, Orion is the first area we would look to for possible variables, and we will not be disappointed! Orion's Great Nebula harbors some of the most fascinating variables of the entire sky. Lurking within the nebula are some 50 variable stars, 10 of which are bright enough to be observed with a 15 cm (6 inch) telescope.

With telescope, winter offers a host of unparalleled, delightful stars that are infrequently observed because of the clouds and the cold. U Geminorum, which can rocket from 14.2 to 8.8 in a few hours, is a highlight of winter observing.

25.1   Observing hints for cold weather

A winter night can be a devastating experience, which under no circumstances should be taken lightly. During his search for trans-Neptunian planets, one quiet, wind-free night, Pluto discoverer Clyde Tombaugh opened the shutter of the 13 inch telescopic camera and began an exposure. He had been out already for some time and looked forward to the chance to sit back and watch the telescope do his work for him. During the slow guiding process, however, he grew tired, feeling strangely comfortable, as though the exposure could go on even longer than he had planned. After the hour Clyde rose to change the film for the next exposure. He could hardly move! With great difficulty he closed the telescope and dome, stumbled inside, and began the painful process of warming up from what he now realized was an early stage of hypothermia. "Had I allowed myself to fall asleep," Tombaugh now admits, "I may never have awakened."

Horror stories like this are distressingly common. My own happened one bitterly cold morning, also somewhat damp, when the telescope tube froze to its mount and I could not separate the two to bring them inside after the session was over. Be extremely careful with cold. Dress warmly, making sure that every part of your body is covered. Be particularly careful of your head, your body's natural escape hatch for warmth. A warm hat is a must for winter observing; in fact, all the clothing in the world is next to useless if your head remains uncovered.

Cold weather offers others hazards; eyes can freeze to eyepieces, and damp breath can fog lenses. Use a portable hair dryer to keep optics dry. Also, keep moving. Observing demands patience and long periods spent motionless beside a telescope, but keeping warm demands motion! Between variable estimates, walk, run, or jump; this will help the circulation, keep your mind alert, and ensure your estimates remain accurate.

25.2 Camelopardalis

In the constellation of the Giraffe (Fig. 25.1) are a few variables that observers in the northern latitudes can observe almost the year round. The most exciting is Z Camelopardalis, leader of that subclass of dwarf novae that stall occasionally on their way down from maximum (see chapter 18).

053068 S Camelopardalis. With a range of over three magnitudes (7.7–11.6), and a period of under a year (326 days) S Camelopardalis does not give you much variation for your observing time. Observe once a month. Semiregular. Level 2.

043065 T Camelopardalis. A red Mira star with a range of almost six

Fig. 25.1. Camelopardalis and Cassiopeia.

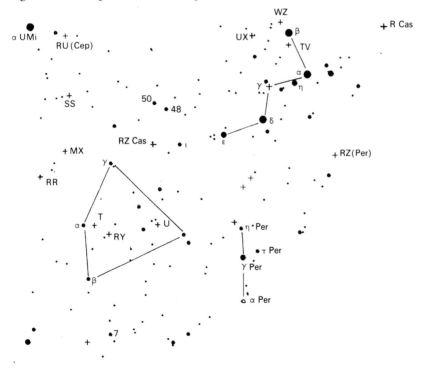

magnitudes, from 8.0 to 13.8. Easily found by moving almost one and a half degrees west of the 4.4 magnitude star Alpha Cam. In 1891, Rev. T. H. Espin, an amateur astronomer with a project to study red stars, discovered the variable nature of T Camelopardalis. As T Cam rises from minimum you may notice two things; about a magnitude before its peak it may simply stop rising for a while (a standstill) or even drop a bit in

Fig. 25.2. Light curve of observations of X Camelopardalis, based on AAVSO and earlier observations, from 1904 to 1964. Reprinted by special permission of the American Association of Variable Star Observers, through its Director, Janet Mattei.

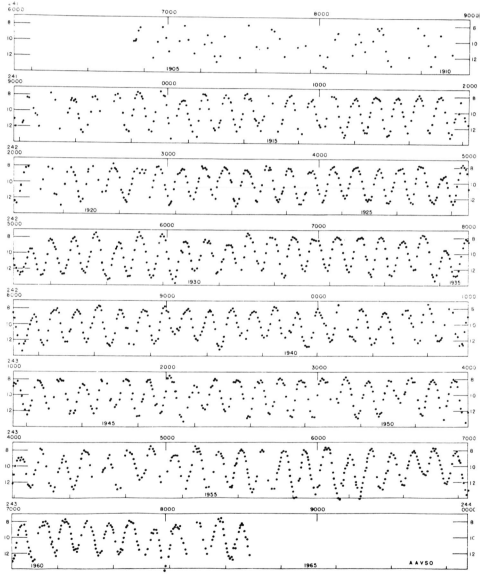

043274   X CAMELOPARDALIS, 1904 – 1964

brightness (a hump). You will see also a leisurely stay near maximum, an event which is typical of stars as red as T Cam. Period 373 days. Level 3.

033362 U Camelopardalis. Semiregular variable, range approximately 11.0–12.8. Be doubly cautious in estimating this very red star. Although its long period of over 400 days suggests that you need not estimate it too often, T often fluctuates with shorter periods. Estimate once every two or three weeks. Level 3.

043274 X Camelopardalis. Mira star, range 8.1–12.6 (Figs. 25.2 and 25.3). If we could watch a variable star all year around, without any seasonal interruptions, we would really be able to get into the star's behavior and know it well. What's more, if our hypothetical variable were to vary fast enough and over a wide enough range, say five magnitudes over five months, then we'd have a star that could present exciting changes every week.

In X Camelopardalis we have found such a star. It fits our dreams entirely and what's more, it is easy to find. The period of 143 days is short enough that you can detect change every week. This star enables you to study the behavior of Mira stars in accelerated time, with two and a half light curves during a single year. Level 2.

071069 RU Camelopardalis. Cepheid variable, range approximately 8.1–9.8, period of 22.06 days. To obtain a good light curve of this star you should estimate every few days. Level 2.

070468 AA Camelopardalis. Irregular variable, range 7.5–8.8. About one degree southwest of RU Cam. Level 2.

053568 AU Camelopardalis. This semiregular variable has a half magnitude range (10.0–10.7) and an unknown period. AU Cam is only a half degree from its better known neighbor S. Level 2.

Fig. 25.3. Light curve of AAVSO observations of X Camelopardalis, from 1975 through 1977. Reprinted by special permission of the American Association of Variable Star Observers, through its Director, Janet Mattei.

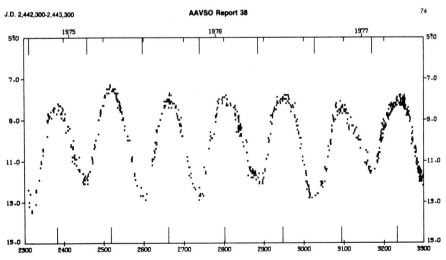

042164 RY Camelopardalis. A semiregular variable, range about 8.0–9.5; period of 136 days. Level 2.

052372 RR Camelopardalis. Semiregular variable, range 9.5–11.3. In a period of 124 days, this star shows sufficient change so that it is worth your while to estimate its magnitude once every two weeks. It is one of the more easily watched stars of the semiregular class. Level 3.

25.3 Perseus

In the midst of Milky Way-studded Perseus lie two interesting long period stars, W and Y Persei. They both remain relatively bright, even at minimum, and thus are easy to observe.

024356 W Persei. Semiregular variable, range 8.7–11.8; period 485 days. W Persei is one of the easiest stars to find; it lies only one degree north of Eta Persei! Moreover, two bright stars of magnitudes 6.5 and 7.1 are practically on top of the variable, and the pattern they form with it and a 9.2 star is very easy to find. Level 2.

032043 Y Persei. With a period of 249 days and a magnitude range from 8.4 to 10.2, Y Persei can be watched and estimated easily. The comparison sequence is ideal, with stars separated in approximately half magnitude increments. Level 1.

031646 RT Persei. A faint eclipsing variable, range 10.5–11.7; period

Fig. 25.4. Perseus.

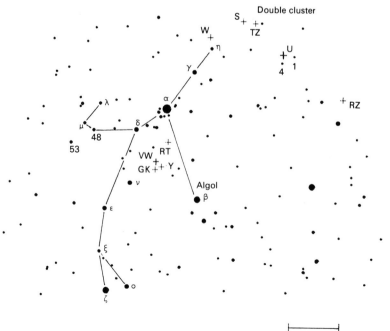

20.4 hours. Its eclipse event lasts four hours, centered on the very sharp minimum. Estimate this star every fifteen minutes, and every five minutes after the star has faded past 11.7, so you will be sure to catch the minimum. In 1904, L. P. Ceraski discovered that this star was variable.

Within a few years the assiduous observer R. S. Dugan of Princeton noticed that the period was becoming shorter, an effect which lasted several decades until the period began to lengthen. Why should this happen? V. A. Dombrowski, a Soviet astronomer, suggested in 1936 that the eclipsing pair might actually be orbiting a third body, and during the half of the orbit near to us, the period would appear shorter. When Dugan disagreed with this theory, he urged other observers to observe this star often. The answer is hidden in some future set of observations from good visual observers. Level 4.

012350 RZ Persei. Mira star, range 9.4–13.7; period 355 days. Level 3.

020657 TZ Persei. Use 5 and 8 Persei as finder stars as both are within a degree of the variable's field. Z Cam star with extremely faint minimum of 15.3; its maxima are at 12.3 and an average of 16.7 days apart, unless it behaves like a Z Cam star and stops declining halfway down. You will need a 40 cm (16 inch) telescope and a bit of patience to find it. Work from the star 5 Persei, and move three-quarters of a degree north and a bit less than a quarter of a degree east. Level 4.

032244 VW Persei, Mira star, range about 11.0–12.5; period 280 days. Estimate this one only once a month. Level 3.

032443 GK Persei: Nova 1901. Discovered as a third magnitude nova by Reverend Thomas Anderson on February 22, 1901, while he was walking, this exciting star reached a bright − 0.5 maximum a day or two later. During its slow decline, the star had many fits and starts, but then it gradually settled down at about 13th magnitude. Now known as GK Persei, this star sleeps fitfully, brightening every year or two by as much as two magnitudes.

In finding GK Per, notice first that both it and Y Persei are east and north of Algol, or Beta Persei. The field is very rich, crowded with many faint Milky Way stars, so you have to be careful to find the correct variable star and its surrounding comparison sequence. Concentrate on the 9.9, the 10.6, and the 10.8 stars, but use also the 9.1 and the 11.8 stars as backups.

GK Per's activity has been observed almost continuously since 1901. According to a 1981 AAVSO "Alert Notice", in the mid 1960s the star brightened suddenly by as much as two magnitudes. It then underwent five minor outbursts between 1966 and 1975. During three of these, GK Per rose suddenly by at least one magnitude, remained at a maximum for a very short time and then declined just as quickly to its 13th magnitude minimum. A fourth outburst showed similar behavior but smaller range, while during a fifth outburst GK Per rose about one magnitude and then declined slowly.

Estimate GK Persei once each night. Level 3.

25.4 Taurus

053326 RR Tauri. From 10.2 to 14.3, this Orion-type variable changes in irregular fashion. Look half a degree north of 125 Tauri. Level 4.

035727 RW Tauri. Located northeast of the Pleiades. If the 2.77-day period of the eclipsing binary RW Tauri is typical of an Algol eclipsing binary, its magnitude range and behavior during eclipse are not. From its 8th magnitude maximum, RW gradually begins its descent, dropping more precipitously as it nears minimum near magnitude 11.6. Level 3.

25.5 Orion

054920 U Orionis. Range 6.3–12.0; period 368 days. Level 2.

050001 W Orionis. Semiregular variable, range 5.9–7.7; period about 212 days. Level 2.

052504 CK Orionis. Semiregular variable, range 5.9–7.1; period approximately 120 days. Bright but has a small range. Level 3.

Fig. 25.5. Taurus.

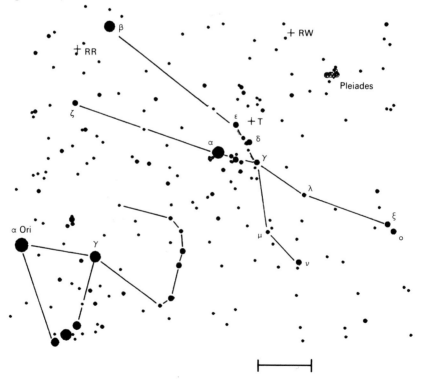

25.6 Lepus

060124 S Leporis. Semiregular variable, range approximately 6.0–7.6; period roughly 89 days. Level 2.

050611 RX Leporis. Irregular variable, range 5.4–7.4. Level 1.

25.7 Auriga

050953 R Aurigae. One of the most popular variables, this long period Mira enjoys a leisurely passage from its 7.7 magnitude maximum to a faint 13.3 and back again in 488 days. R is not near any bright stars, but it is almost due north of Capella by about seven and a half degrees. A fairly red star, you should be careful not to glance too long at it lest your eyes fool you by brightening it up too much. You can estimate this Mira star once every two weeks. At its maximum R can be seen without

Fig. 25.6. Orion.

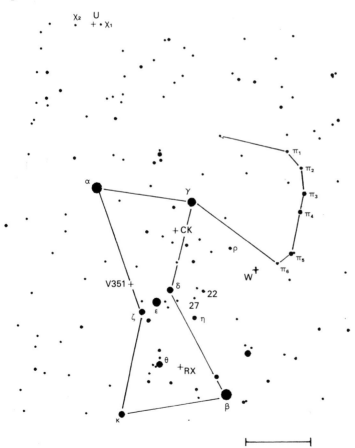

difficulty in a 10 cm (4 inch) telescope, but as its light level drops, you need a larger instrument. Level 2.

052034 S Aurigae. Semiregular variable, range 8.2–13.3. The period is 590 days but according to Kukarkin's *General Catalogue of Variable Stars*, S Aur stayed faint with only minor variations during the early 20th century. Level 2.

061647 V Aurigae. A bit more than one and a half degrees south of Psi Aurigae, V Aur varies from 9.2 to 12.1 and back in a period of just under a year, 353 days. The field is somewhat sparse, and so is the comparison sequence. Level 4.

060450 X Aurigae. Mira star, range 8.6–12.7; period 164 days. Found via 41 Aurigae, a 5.6 magnitude star. The field of X is about 1.5 degrees north and slightly east. Level 2.

060443 RR Aurigae. Mira star, range 9.6–13.6; period 308 days. Level 3.

062230 RT Aurigae. A bright Cepheid with a period of 3.73 days, this star was discovered as variable in March 1905 by T. H. Astbury during a search for novae. Its range from 5.0 to 5.8 is slight but its variability so rapid that you should have no trouble observing this star with binoculars. Its type of variation is classical Cepheid, with a faster rise than fall. Level 3.

050130 RW Aurigae. A rapid irregular variable with a range of approximately 9–13. This star can vary rapidly, featuring quick changes

Fig. 25.7. Lepus.

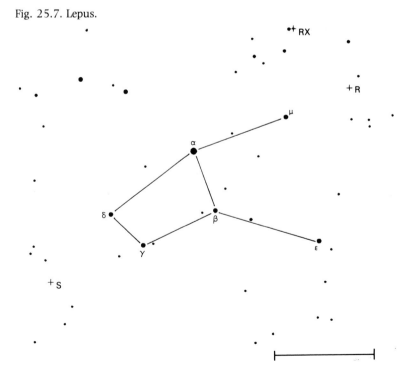

of two or more magnitudes seen over a few hours. Estimate once per night. Level 4.

060547 SS Aurigae. An unusually slow dwarf nova with a period of about 558 days between outbursts, SS Aurigae normally hovers around 15.8 magnitude. During outburst it rises to 10.3. Level 5.

062938 UU Aurigae. A very red semiregular variable star with a range of over one and a half magnitudes and a complex cycle. Two different periods are superimposed on each other, one being about 9.3 years (3500 days), and the other lasting a comparatively short 234 days. Since both periods are long, you need estimate UU Aur only once a month. You can find UU Aurigae by going northwest about 12 degrees from Castor (Alpha Geminorum). Level 2.

045443 Epsilon Aurigae. Eclipsing binary. Range 2.9–3.8. The period of this strange eclipsing pair is 9899 days, or slightly over 27 years. The decline lasts 197 days, minimum lasts about a year (360 days), and the rise takes another 197 days. The first minimum was observed in 1821, and three more were observed in 1847–8, 1874–5, and 1901–2, before the eclipsing nature of the variation was confirmed. Since then the star has undergone eclipses in 1929, 1956, and 1983–4. Level 1.

Fig. 25.8. Auriga.

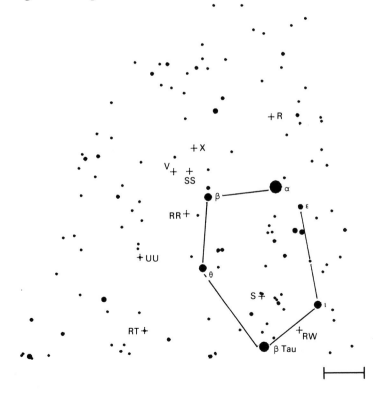

25.8 Canis Major

071416 R Canis Majoris. Dropping every 27 hours from 5.7 to 6.3, this eclipsing binary was the first variable to be discovered in Canis Major, by amateur astronomer Edwin F. Sawyer in 1887. The shape of the curve, especially that of minimum, has changed since 1920. Level 1.

070311 W Canis Majoris varies from 6.3 to 7.9 without any regularity. As the changes it shows are not rapid, you need estimate it only once a month. Level 2.

061317 UY Canis Majoris. RV Tauri star, somewhat similar to R Scuti. Range approximately 10–12; period 114 days. Needs more observations. Level 3.

071825 VY Canis Majoris, an 8th magnitude red irregular variable. Actually this "star" consists of a nebula in which a number of starry condensations appear. These cannot be seen separately in amateur telescopes, but the condensations as well as the nebula seem to vary.

This unusual behavior were first noticed by Leif Robinson and reported in *Sky and Telescope*, November 1970 issue. Range 6.5–9.6. Level 4.

Fig. 25.9. Canis Major.

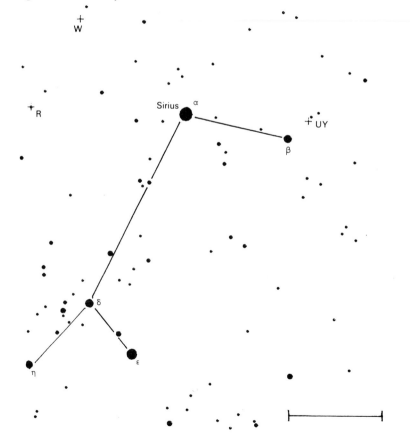

25.9 Gemini

060822 Eta Geminorum. Bright semiregular star with a range 3.2–3.9, and a 233 day period. Eta Geminorum is a bright 3rd magnitude star whose variation does not even require binoculars to see. If you observe Eta Gem once every week, you may begin to understand what it does after a season. I use the word "begin" on purpose, for other observing seasons may show slightly different patterns of variation. Your friendship with this star, as with all, needs more than a year to mature. Level 1.

065820 Zeta Geminorum is a Cepheid variable whose 10 day period is slightly longer than typical. What is also not typical is the shape of its light curve; while many classical Cepheids present rises that are faster than their falls, Zeta Geminorum brightens at about the same rate as it later declines. Range 3.6–4.2.

In 1844 the amateur astronomer J. F. J. Schmidt thought that Zeta Gem was not constant in brightness. According to Joseph Ashbrook, he later confirmed this while an undergraduate at Bonn University in 1847. Level 3.

070122 R Geminorum. Mira star, range 7.1–13.5; period 370 days. This star is fun for three reasons. Its range is large, so that if you observe it every two weeks you should see a lot of change with each sighting. Its

Fig. 25.10. Gemini.

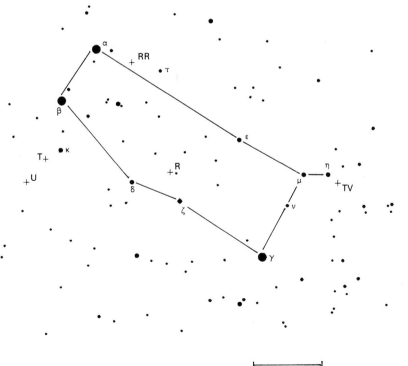

period of 370 days is so close to a year that you can almost plan your birthday by a particular magnitude it reaches as it rises or falls. Further, it is less than three degrees north and a bit east from Zeta Geminorum, and about half a degree east from 6th magnitude 44 Geminorum. Level 2.

074323 T Geminorum. Mira star, range 8.7–14.0; period 288 days. Level 3.

071531 RR Geminorum. Varies between magnitudes 10.6 and 12.0. This cluster variable rises to maximum within an hour, only then to spend more than eight hours fading back to minimum. Although the period has varied in the past, it has been constant over a number of years. A good reason for observing a Cepheid variable is to pin down a possibly changing period. With RR Gem, the period may be lengthening slightly. Shortly after the nature of this star was discovered in 1903 by L. P. Ceraski of Moscow University, its period of 9.5 hours was established. The rising part of the curve is exciting to watch, since RR leaps from 11.9 to 10.7 in just over an hour. Level 2.

060521 TV Geminorum. Semiregular variable, range approximately 8–9; period 182 days. Level 2.

25.10 Monoceros

061907 T Monocerotis. Discovered as variable in 1871 by B. A. Gould, this Cepheid varies from 5.6 to 6.6 in 27.02 days. Since 1900 the period has been constant, although at that time it abruptly increased by 13 minutes. Difficult to observe visually with accuracy. Level 3.

072609 U Monocerotis. A field containing several interesting bright

Fig. 25.11. Monoceros.

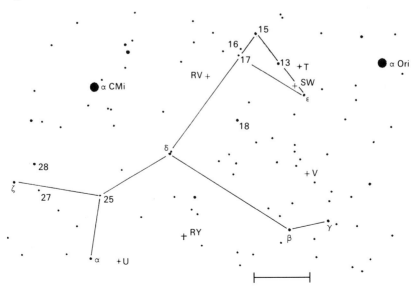

variables. U Monocerotis is a 91 day RV Tauri variable with a maximum of 5.6 and a minimum of 7.3. It is not as red as most of the bright variables. It is somewhat similar to R Scuti, except that the 92 day period is superimposed on a much longer period of about 2320 days. Level 2.

061702 V Monocerotis. Mira star, range 7.0–13.1; period 340 days. Level 2.

070207 RY Monocerotis. Semiregular variable, range 7.7–9.2; period 456 days. Level 2.

062105 SW Monocerotis. Semiregular variable, range 9.0–10.9; period 112 days. Level 2.

25.11 Puppis

071044 L2 Puppis. This bright variable is visible only from more southerly latitudes. This red star lies in a beautifully rich field, studded

Fig. 25.12. Puppis.

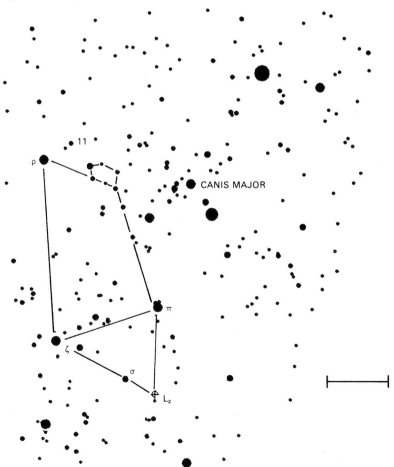

with moderately bright stars. L2 Pup is 180 light years away. Its maximum can be as bright as 2.6, and its minimum is 6.2. Since the star varies slowly over a period of about 141 days, you do not need to estimate it more often than once every two weeks. Its red color makes this semiregular star quite a treat, but watch for Purkinje effect. Level 1.

26   April, May, June

As this season opens, the excitement of Orion is now at our disposal in the early evening. R Leonis is a fine star, and then you can move your thoughts and your telescope to RY Leonis, a long period star that is easy to find.

Next, move over to the Big Dipper region, to a small region that houses the "Dipper Trio". It consists of two long period variables, T and RS Ursae Majoris, with closely matched periods, and S UMa with a shorter period but altogether a brighter star. Spring also offers perhaps the most unusual dwarf nova of all, Z Camelopardalis, a star whose exploding fluctuations are often interrupted by standstills.

26.1   Lynx

072046 Y Lyncis. Red and bright, this semiregular has a half magnitude range from 7.2 to 7.8 and a possible but imperfectly defined period of about 110 days. To find this star with your binoculars, go about 13 degrees east of Beta Aurigae. Difficult to follow usefully because of its small range. Observe it only once a month, no Moon, constant observing conditions from one observation to another, same pair of binoculars, and watch out for the Purkinje effect. The more you keep observing conditions constant, the more accurate your Y Lyn picture will be. Level 2.

061359 U Lyncis. Mira star. Range 9.5–14.4; period 436 days. Level 3.

26.2   Cancer

081112 R Cancri. About 10 degrees southwest from the Beehive or Praesepe Cluster, M44, lies a bright Mira variable with a period of about a year (362 days) and a range of over four magnitudes. Its maximum of 6.8 is bright enough to be visible through binoculars, and the minimum of 11.2 is easily observable using a small telescope. To find R Cnc, go about two and a half degrees north of the 4th magnitude star Beta Cancri. Level 2.

085020 T Cancri. Semiregular, range 7.6–10.5; period 482 days. Level 2.

084917 X Cancri. One of the best of the bright variables of northern spring, this star is located southeast of M44, the Beehive cluster. Its range of 5.6–7.5 keeps it bright all the time, easily followed with binoculars, but its period of 165 days is slow, encouraging observations but once every month. Level 1.

085211 RT Cancri. Range 7.1–8.6. This variable lies about two degrees south of Alpha Cancri, and is also southeast of the beautiful cluster M67. A red star, this semiregular's behavior suggests a rather short period of about 90 days; thus you would be wise to observe this star about once each week. This presents a problem, however, since red semiregulars need constant conditions for accurate results in observing. If you make four estimates each month, the changing phases of the Moon will make these conditions hard to satisfy. If you can, observe RT Cnc when the Moon is not in the sky. As this will not always be possible, at least record with your estimate whether you think that the Moon is interfering. If you observe RT Cancri for a long period of time, you could possibly notice a slight variation of the mean magnitude (a magnitude between the maximum and minimum brightness); it does vary over a period of 540 days. Level 2.

Fig. 26.1. Lynx.

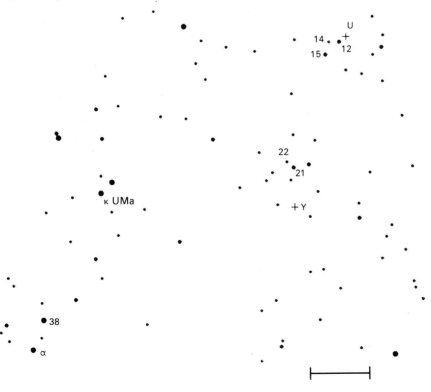

085518 SY Cancri. Z Cam star, range 10.6–14.0; questionable period of about 27 days. Level 4.

080428 YZ Cancri. A dwarf nova with an extremely faint photographic minimum of 14.6 and a high of 10.6. However, it appears to be very active, with a period as short as 11 days. Needs a 40 cm (16 inch) telescope and good dark sky for a view throughout the cycle. Level 4.

Fig. 26.2. Cancer.

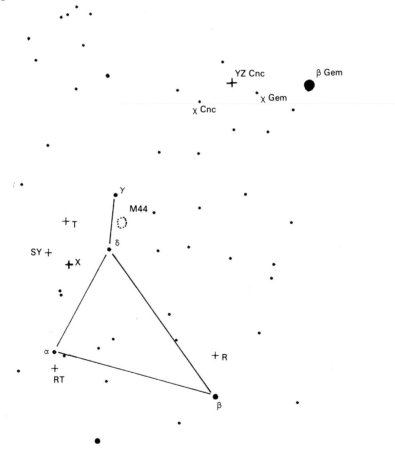

26.3 Ursa Major and vicinity

103769 R Ursae Majoris was discovered as variable by Norman Pogson from Oxford, England. Pogson and his observing group were well known for their important work during the mid 19th century. Noticing some inconsistency in R UMa's light during a decade of observations, Pogson confirmed their work with the announcement of the first variable in the Big Dipper region. Range 7.5–13.0; period 302 days. Level 2.

The Dipper trio

123961 S Ursae Majoris. Range 7.8–11.7; period 226 days. Level 2.
123160 T Ursae Majoris. Range 7.7–12.9; period 257 days. Level 3.
123459 RS Ursae Majoris. Range 9.0–14.3; period 259 days. Level 4.
I have always had a special feeling for these three stars in Ursa Major. In a telescope of sufficiently wide field an observer can see all of them at the same time. As well, from northern latitudes these stars are observable all year round. But aside from that, these are "average" Mira stars, and while bright S is fairly easy, T and RS are quite difficult when faint.

Fig. 26.3. Ursa Major.

Like an out-of-tune singing group, these three Mira stars seem to perform to the mysterious beat of their own drummer. Each is a long period variable. Notice that the periods of T (257 days) and nearby RS (259 days) are only two days off which means that they appear to vary in synchronization with one another. S UMa, with a somewhat shorter period (226 days), appears to be somewhat off, like the penultimate chord of a Bach fugue. Of course, the symphonic analogy is coincidental, but as observers, we look for ways to add rhythm to our work, and even though the musical theme is illusory, it does add a certain zest to our view of the stars in this corner of the Dipper.

080362 SU Ursae Majoris. U Gem star, range 10.8–15.0; period averages 17 days but like most U Gems, varies considerably. Representative of a strange subclass of U Gem stars, this star displays what we call a supermaximum every few outbursts, during which the star's maximum brightness is greater than usual and it stays in outburst longer. Further, oscillation events called "superhumps" averaging 2.4 magnitudes occur during these interesting times.

Fig. 26.4. Ursa Major and Lynx.

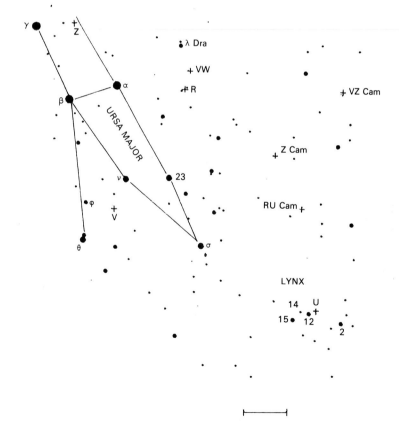

SU UMa occasionally shows rapid, irregular flickering while at minimum. The star has brightened and faded by up to three-quarters of a magnitude in a five minute interval. Level 5.

103946 TX Ursae Majoris. This Algol-type eclipser ranges from 7.0 to 8.8 in a period of 3.06 days. The period of TX UMa changes over a cycle of 32 years. Level 2.

093156 VV Ursae Majoris. This fast Algol star varies by a magnitude, 10.1 to 10.9, over 16.5 hours with a time for main eclipse lasting just over 3 hours. First observed by Helen Gitz on Moscow photographic plates. Level 2.

Northern Dipper group

105270 VW Ursae Majoris. Range 7.2–7.8; period 125 days. Level 2.

121561 RY Ursae Majoris. Range 7.0–8.0; period 311 days. Level 2.

125266 RY Draconis. Range 6.0–8.0; period 200? days. Level 2.

For observers with binoculars, these three semiregular stars provide occasional activity. Because they are so far north, they are especially convenient for northern hemisphere observers, and can be watched much of the year.

105270 VW Ursae Majoris presents two full cycles per observing season for observers in north-temperate latitudes. Although its normal range is from 7.2 to 7.8, I have seen it drop to 8.2. In any case, the range is not sufficient to make visual work reliable unless you are careful to keep your observations as consistent as possible. Always observe at the same lunar phase (preferably with no Moon in the sky at all), and with the sky free of clouds.

Although 121561 RY Ursae Majoris has a range of one full magnitude, the period at 311 days is long. The star is not difficult to find, being less than five degrees north of Delta Ursae Minoris and near the long period variable T UMa. Because its period is so much longer, the variation from observation to observation is less, so make sure you do not estimate this star more often than once a month.

125266 RY Draconis is the most interesting of these three. Its 200 day period is highly uncertain. This star is something of a challenge for visual observers. Carefully watched, this star could divulge some of its secrets to visual observers. During my observing, I did not detect much more than a 6.7–7.3 variation, over "cycles" ranging from three to eight months. Observe once every two weeks.

Southern Dipper group

112245 ST Ursae Majoris. Range 6.0–7.6; period 81 days. Level 1.

124045 Y Canum Venaticorum. Range 5.2–6.6; period 157 days but this is difficult to detect. Level 3.

125047 TU Canum Venaticorum. Range 5.5–6.6; Irregular. Level 3.

131546 V Canum Venaticorum. Range 6.5–8.6; period 192 days. Level 3.

These four bright semiregular stars are all visible in binoculars. We begin with 112245 ST Ursae Majoris, one of the best semiregulars for observing. ST offers a substantial variation for a bright red supergiant, as well as a period of only 81 days. Moreover, the comparison sequence is unusually good, with easily found stars well placed near the variable. Enjoy it!

124045 Y Canum Venaticorum. Named "La Superba" by the 19th century observer Angelo Secchi, this star is so red that visual estimates are quite difficult.

125047 TU Canum Venaticorum. An irregular variable with about a magnitude range, from about 5.5 to 6.6, this star also needs to be watched with care, lest your observational scatter exceed the range of the variable!

131546 V Canum Venaticorum. In the April 1971 *Journal of the Royal Astronomical Society of Canada*, the AAVSO reported that V CVn "has

Fig. 26.5. Canes Venatici.

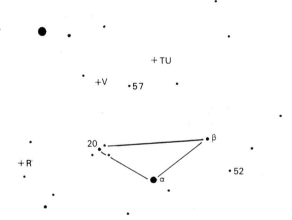

become too irregular to predict, and the range has decreased to about half a magnitude." In 1976, however, the same journal reported that V CVn had "resumed periodicity in 1971." Once again, be careful!

26.4 Ursa Minor

163172 R Ursae Minoris. Semiregular, range 8.5–11.5; period 326 days. Can be followed year round by most northern hemisphere observers. Level 2.

153378 S Ursae Minoris. Because this star is so close to the north celestial pole and can be watched all year from most northern observing sites, it is one of the more interesting Miras. The period is 331 days, about 11 months, and the range of 8.4–12.0 offers easy observing with even a 15 cm (6 inch) telescope. S UMi is just over a degree north of the 5.3 magnitude star Theta Ursae Minoris, one of the stars in the Little Dipper's bowl. Observe every 2 weeks. Level 2.

A 15 cm (6 inch) telescope is sufficient for S. Theta and Xi Ursae Minoris are good guide stars that make it easy for you to find S.

Fig. 26.6. Ursa Minor and Draco.

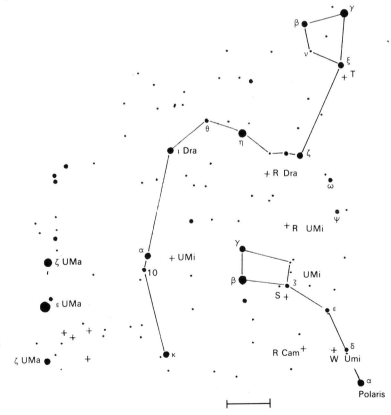

141567 U Ursae Minoris. Mira star, range 8.2–12.0; period 331 days. Level 2.

163486 W Ursae Minoris. Near the 4th magnitude star Delta Ursae Minoris is a fine eclipsing binary whose nature was discovered in 1913 by two British astronomers, an amateur named T. Astbury and a professional at Greenwich Observatory, M. Davidson. Called W UMi since it was the sixth star to be recognized as variable in Ursa Minor, this star varies over 40.8 hours. The eclipses last about ten hours, evenly spaced around minimum. The period has been known to change suddenly by small amounts. Range 8.5–9.6. Level 3.

26.5 Leo

110506 S Leonis. Mira star, range 10.1–13.9; period 190 days. Level 3.

104814 W Leonis. Mira star, range 9.8–14.2; period 392 days. Level 3.

094512 X Leonis. This U Geminorum-type dwarf nova is one of the best for the spring sky. Extremely easy to find, it practically touches the 6.6 magnitude star 21 Leonis, and it is just one degree east (but slightly north) of the famous long period variable R Leonis. The problem develops during the estimating process, for at minimum the star is about 15.7 magnitude. The faint minimum does not necessarily mean that you cannot observe it with a small telescope, for during outburst it rises to 11.1, a magnitude well within the grasp of a 20 cm (8 inch) telescope. The behavior pattern of these stars is usually all-or-nothing, so if the star has been much below maximum for some time it likely is pretty close to its minimum. Thus, a nightly observation program of X Leonis is possible with a 20 cm (8 inch) or 30 cm (12 inch) telescope, so long as you don't mind many nights when you must record merely the magnitude of the faintest star visible. If the faintest star you see is 13.3, and the variable is invisible, then you simply record (13.3, meaning fainter than 13.3. Period about 17 days. Level 5.

093126 Y Leonis. One of the most dramatic eclipsing variables around! Y Leonis at maximum is an easily found star of magnitude 10.0, but once the eclipse begins the star fades fast, falling in three hours all the way to 13.2 magnitude. After twenty minutes at minimum, the star begins a corresponding three-hour rise to maximum again. The changes are so fast that you can estimate this star every five or ten minutes and see substantial change. The period changes slightly. Level 3.

100224 RR Leonis is known as a "cluster" variable. Discovered by Henrietta Leavitt during her variable star work at Harvard, this bright variable passes through a complete cycle every 11 hours, during which it will drop slowly from its 9.9 maximum to 11.3. The rise that follows is spectacular, happening in only 80 minutes! RR Leonis is slightly west of Zeta Leonis, inside Leo's sickle. Period changes slightly. Level 3.

093720 RS Leonis. A faint Mira with five and a half magnitudes of variation. Its faint 10.7 maximum is bright enough for a 20 cm (8 inch) telescope, but at minimum it is around 16.0, needing 40 cm (16 inches) or more for most observers. Its 208 day period is a bit short for typical Mira stars. Estimate once every two weeks. Level 4.

095814 RY Leonis. Just two degrees south and one degree west of Regulus, RY Leonis is an excellent star for beginners. Its range of three magnitudes (9.0–11.8) makes it bright enough to be followed with a 20 cm (8 inch) telescope, and its period of 155 days make it ideal for observers who wish to obtain a full light curve in a single observing season. Level 3.

26.6 Leo Minor

093934 R Leonis Minoris. Mira star, range 7.1–12.6; period 372 days. R Leonis Minoris demands just a little more of your time and patience than does its almost-namesake, R Leonis, because it is slightly harder to find and its minimum is fainter. Level 2.

Fig. 26.7. Leo.

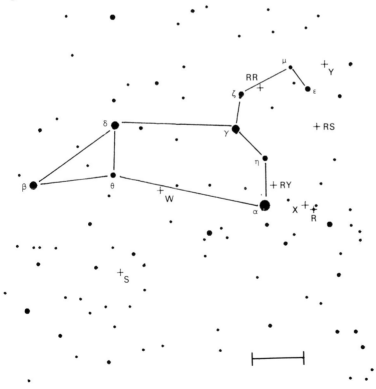

094735 S Leonis Minoris. Mira star, range 8.6–13.9; period 234 days. Easily followed once you find it, but since there are no nearby bright stars, finding the star may be a challenge. Level 3.

094836 U Leonis Minoris. Mira star, range 10.0–13.3; period 272 days, semiregular. Level 4.

26.7　Virgo

123307 R Virginis. Mira star, range 6.9–11.5. The first discovered variable in Virgo, R Virginis is a good beginner's star because it is fairly bright even at its 11.5 magnitude minimum. Also, its 145 day period is short enough that you should see considerable change each time you observe the star, even if your observations are spaced as close as a week apart. If it is near maximum you may see it with binoculars or the finder of your telescope. If not, try finding the star by turning to the other stars that can more easily be seen through your telescope. Try the 5.5 magnitude star called 31 Vir, and 5.2 magnitude 32 Vir which are close by. Level 2.

124606 U Virginis. Mira star, range 8.2–13.1; period 207 days. Level 2.

Fig. 26.8. Leo Minor.

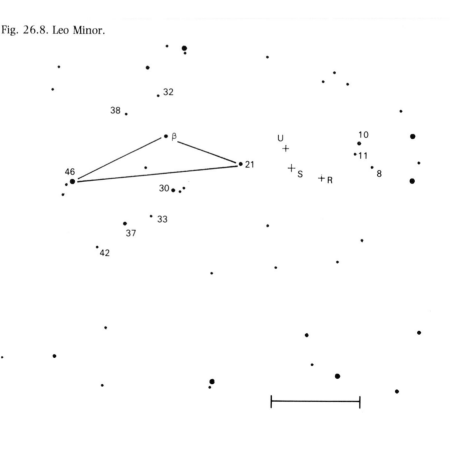

122001 SS Virginis. If you look at R Vir, you should also examine nearby SS Virginis once a month. A slightly semiregular variable, SS Vir has a bright maximum of 6.0 and a minimum of 9.6, and a period of just a few days short of a year. The only difficulty with this star is that its comparison star sequence uses stars separated by considerable distances as seen through your telescope. Level 3.

26.9 Canes Venatici

134440 R Canum Venaticorum. Mira star, range 7.3–12.9; period 329 days. Level 2.

26.8 Hydra

132422 R Hydrae. One of the earliest known variables, this Mira star's light changes were discovered by the 17th century Italian observer Geminiano Montanari. Its period during those early observations was about a year and a half, but in a most unusual long term change, the cycle has shortened to just over a year, at 389 days. Range 3.5–10.9. Level 2.

085008 T Hydrae. This star's behavior is fairly close to what is "average" for Mira stars. Its 298 day period and its five magnitude range (7.8–12.6) make it a good beginner's star. Also (and this is important for a beginner's star) it is easily found by moving south just over a degree from 6th magnitude 17 Hydrae. Level 1.

Fig. 26.9. Virgo.

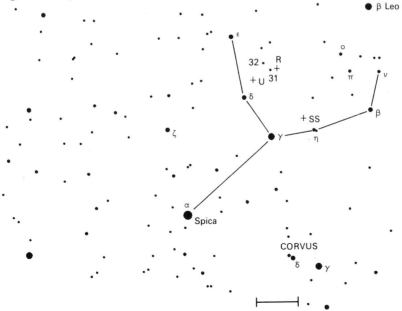

104620 V Hydrae. I met V Hya surreptitiously one morning while searching for comets. Field after field went by, and then, there appeared one of the reddest stars I have ever seen. This is a star you do not have to work with to appreciate; just a look at it and its field is a thrill. According to the *General Catalogue of Variable Stars*, the "mean" or average magnitude changes over 18 years by about 5 magnitudes, so that deep minima were seen in 1891, 1909, 1926, 1942–3 and 1959–62. Period 530 days. Level 3.

082405 RT Hydrae. Range 7.0–10.2; period 290 days but semiregular. Level 2.

124728 EX Hydrae. Range about 9.6–14.0. A U Gem star that, like SU UMa, exhibits occasional "supermaxima." Also an eclipsing binary. Level 5.

26.10 Bootes

143227 R Bootis was discovered to be variable by E. Schonfeld and F. W. Argelander from Bonn Observatory during the preparation of the mid 19th century atlas and catalog, the *Bonner Durchmusterung*. Range 7.2–12.3, period 223 days. Level 2.

142539 V Bootis. Only one degree northwest of third magnitude Gamma Bootis, this star is easy to find. Its magnitude normally varies from 7.0 to 12.0 in 258 days, but in the early 1970s this star simply stopped varying for an extended period, only to resume afterwards. Bright, easily found, and quite interesting! Here is a star of spring whose bright minimum means that you can follow its whole period with a 10 cm (4 inch) reflector. Level 1.

135126 ZZ Bootis. Between Arcturus and the famous globular cluster M3 lies this fine Algol star. Its period is almost exactly five days, which means that observers can catch minimum after minimum during night hours, or conversely, miss every one of them!

Fig. 26.10. Hydra.

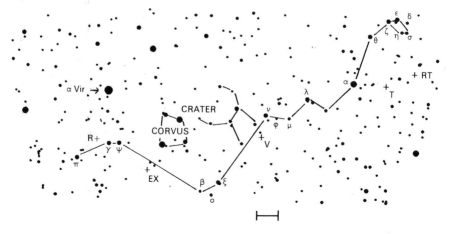

ZZ's maximum is magnitude 6.7. It drops to 7.4 during both its primary and secondary minima. The time below maximum light takes about six hours, roughly three each for fall and rise.

The nature of ZZ Bootis was revealed in 1951 by Sergei Gaposchkin, of Harvard University. Level 2.

26.11 Draco

See Fig. 26.6 for a finder chart.

163266 R Draconis. Mira star, range 7.6–12.4; period 246 days. Slightly over two degrees north and and just less than one degree west of 5th magnitude 18 Draconis. Level 2.

175458a. T Draconis. This star is slightly less than one and a half degrees north of 4th magnitude Xi Draconis. A Mira star, it ranges from

Fig. 26.11. Bootes.

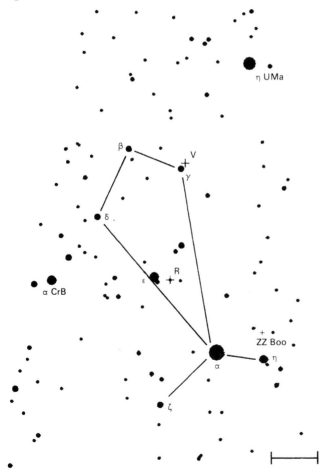

9.6 to 12.3, with a very long period of 422 days. A close neighbor of T Dra is 10.7 magnitude UY Draconis, a star with which T can be easily confused if you are not careful. UY Draconis is a suspected variable of very low amplitude. Level 3.

190967 U Draconis. Mira star, range 9.5–13.8; period 316 days. Level 3.

195277 AB Draconis. Z Camelopardalis star. From a minimum of fainter than 15.3, it rises to a 11.0 magnitude maximum. Period ranges from 8 to 22 days; outbursts last from 2.5 to 7 days. Level 5.

26.12 Libra

151520 S Librae. Mira star with a short period of 192 days, a bit over six months. It ranges from 8.4 to 12.0. Find the star by moving a little more than one degree northeast of the globular star cluster NGC 5897, or a bit over 2 degrees northeast from the pair of stars 1 and 25 Librae. Level 2.

153620a U Librae. Mira star, range 9.6–14.4; period 227 days. Quite faint at its minimum, this star needs a 30 cm (12 inch) telescope to be followed throughout its 5 magnitudes of variation. Level 4.

Fig. 26.12. Libra.

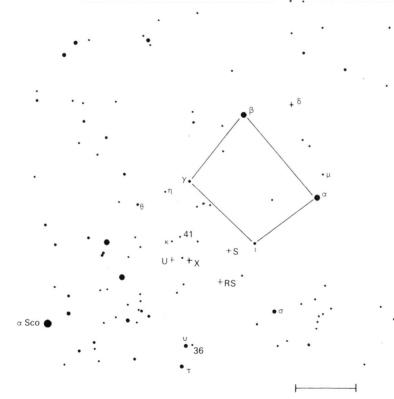

153020 X Librae is a red Mira star with a faster period, 164 days, than nearby U Lib. It ranges from 11.0 to 13.5. A bit less than two degrees almost due north of 5th magnitude Kappa Librae. Level 4.

151822 RS Librae. Almost five magnitudes of change, from 7.6 to 12.3. The period of 218 days is short for a Mira star, but you could still estimate once every two weeks. Level 3.

145508 Delta Librae. A bright eclipser similar to Algol, this star fades by about a magnitude every 2.327 days. The maximum of 4.9 is easily visible through binoculars, or unaided eye from a dark country sky, and its 5.9 minimum should also present little problem. Level 1.

Delta Librae's variability was discovered by the astronomer J. F. J. Schmidt in 1859, from Athens. Like Algol, this is a good beginner's eclipsing star.

26.13 Lupus

155640 EX Lupi, a nebular variable, irregular, with a visual range of 8.7 to 14.2. Level 3.

155240 RY Lupi, an irregular Orion-type variable with a range of

Fig. 26.13. Lupus.

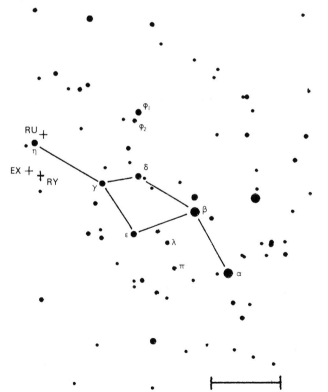

9.7–11.7. Close to EX Lupi. Both stars are about two degrees south of the 3.6 magnitude star Eta Lupi. Level 3.

27 July, August, September

The northern hemisphere nights at this time of year are short but busy, for there is always something to do.

We begin with the most prominent feature of the northern sky, the "summer triangle" of Vega, Deneb, and Altair. In and around this triangle are many fine variables, including SS Cygni, a dwarf nova whose bimonthly explosions have made it one of the most popular variable stars of all.

A good star with which to begin is Beta Lyrae, a star offering nightly variation, and the parallelogram of Lyra offers two additional stars that work well as comparison standards. Next you can move to Aquila, whose Eta provides changes over a slightly longer period. We then can check on our old friends X and g Herculis and RR Coronae Borealis.

At this time of year we can turn our sights to Corona Borealis, the crown of Ariadne where our friends R and T Coronae Borealis lurk in the evening sky. We also can enjoy some of the Crown's more traditional long period variables, like W, whose striking red color helps in finding.

Cygnus offers so many variables that you could hardly begin to see them in a single season. The constellation's most famous variable is Chi, a star lying in the middle of the Swan's neck. This season also features Sagittarius, whose collection of over 2500 variable stars surely must contain something of interest to a beginner. And there is: RY Sgr is a star like R Cor Bor.

This season is special. With its meteor showers, it offers some fine nights for meteor observing, a chance to enjoy the sky without a telescope. Northern summer also provides the most difficulty with mosquitoes, so you need to remember to bring lots of insect repellent for your protection. You should remember also not to apply the repellent to your telescope. Mosquitoes prefer you to your telescope anyway, and any plastic on your telescope may react unfavorably to insect repellent. Welcome to a season of contrasts both in the sky of variables and on Earth.

27.1 Corona Borealis

151731 S Coronae Borealis. Mira star, range 7.3–12.9; period 360 days. Level 2.

151432 U Coronae Borealis. Drops from 7.7 to 8.8 in five hours, then takes the same time to recover. An eclipsing binary with a period of 3 days 10.9 hours. Period changes very slightly but irregularly. Level 3.

154539 V Coronae Borealis. A Mira variable, period 358 days; range 7.5–11.0. If constellation boundaries were real things, lit up in the sky as they so often are on charts, this variable would be the easiest in the sky to find. It lies practically on the mutual boundary of three constellations, Corona Borealis, Bootes, and Hercules. Also, V whimsically stands for *very* red! Be careful! It is one of the reddest stars in the entire sky. Level 3.

161138 W Coronae Borealis. From 5th magnitude Tau CrB, travel northeast almost a degree and a half. W CrB is a Mira variable with a period of 238 days and an range of 8.5 to 13.5. Level 3.

27.2 Serpens

154615 R Serpentis. Mira variable detected by K. L. Harding in 1826 during the preparation of a star chart. Located almost midway between Beta and Gamma, so it is easy to find if it is near its 6.9 magnitude maximum. Can get as faint as 14th magnitude, although its usual minimum is about 13.4. Period 356 days. Level 4.

151614 S Serpentis. Mira star, range 8.7–13.5; period 371 days. Level 4.

Fig. 27.1. Corona Borealis.

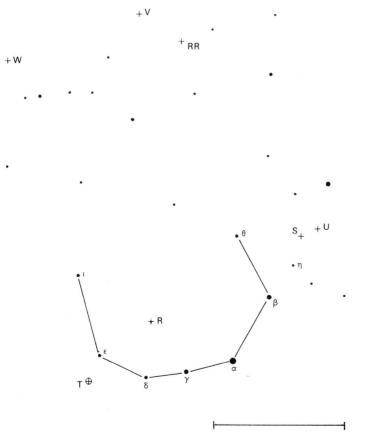

27.3 Ophiuchus

   170215 R Ophiuchi. Typical of first confirmed variables in a constellation, R Oph is a Mira star with a bright maximum (in this case 7.6) and a much fainter minimum (13.3). Such stars were easy to discover because of their alternating periods of visibility and invisibility. In this case, R Oph has help in that it is extremely easy to find, lying about three-quarters of a degree southwest of the 2.6 magnitude star Eta Ophiuchi. During an "average" cycle, this star drops more than six magnitudes and then climbs back up in a period of 306.5 days. Level 3.

   162112 V Ophiuchi. A 298 day period and a shallow range of less than three magnitudes characterize this unusual Mira star. It is also quite red. Range 8–11. Level 3.

   183308 X Ophiuchi. Period 329 days, but semiregular. Range 6.8–8.8. Estimate once per month. Level 3.

   174406 RS Ophiuchi. This is a recurrent nova, a bit like T Coronae Borealis except that it goes into outburst more frequently. Its 1985 outburst was discovered by Warren Morrison, an active amateur observer from Ontario, Canada. Other outbursts occurred in 1898, 1933, 1958, and 1967. Its normal minimum is 11.8, and its maximum ranges from 4.3 to about 6. Even at minimum, RS Ophiuchi goes through changes as much as a full magnitude. Find this star by moving about two and a half degrees southwest of 4.6 magnitude Zeta Serpentis. Level 3.

   172809 RU Ophiuchi. An easily found Mira star, less than 1/2 degree west and a tiny bit south of 6th magnitude 53 Ophiuchi. Period of almost

Fig. 27.2. Serpens.

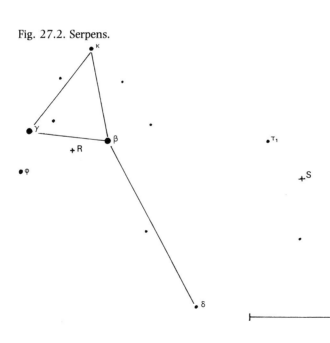

seven months (202 days) so that you may be able to get a complete light curve in a single season of observing if you begin while the star rises in the morning sky. Range 9.3–13.8. Level 3.

181103 RY Ophiuchi. Period 150 days; magnitude range 8.2–13.2. This Mira variable can be located one degree west and a quarter degree north of the 5th magnitude star 74 Ophiuchi. Its period of about five months should allow you to complete a light curve in a single season if you begin early enough. Level 3.

164403 TT Ophiuchi. An RV Tauri star in the same class as R Scuti. A variable of unusually short period, TT Ophiuchi varies by one and a half magnitudes, from 9.5 to 10.8, in only 61 days. Its variation can be followed in a 15 cm (6 inch) reflector. Estimate once every week. Level 2.

165926 BF Ophiuchi. Located just one degree southeast of Messier 19, a fine globular cluster, this bright Cepheid varies over 4.07 days between 6.9 and 7.7. Estimate once per night. Level 2.

Fig. 27.3. Ophiuchus.

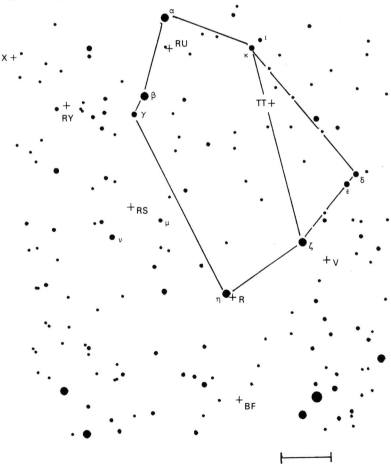

27.4 Hercules

171014 Alpha Herculis. Semiregular, varies from 2.7th to 4th magnitude over two superimposed periods; one lasts six years, the other lasts from 50 to 130 days. Very red star. Level 1.

164715 S Herculis. Mira star, range 7.6–12.6; period 308 days. Level 2.

180531 T Herculis. Mira star, range 8.0–12.8; period 165 days. Level 2.

162119 U Herculis. Mira star, range 7.5–12.5; period 405 days. Level 2.

163137 W Herculis. Mira star, range 8.3–13.5; period 280 days. This is a star designed for those who dislike variables, because for each observation you get a bonus look at M13, one of the most magnificent globular clusters in the entire sky. The field of W Herculis is less than two degrees west, and less than one degree north, of the cluster. Level 3.

171723 RS Herculis. Mira star, range 7.9–12.5; period 220 days. Level 2.

160625 RU Herculis is a slowly varying (485 days) long period Mira star with a 6 magnitude range of 7.9–13.7. This star can also be fruitfully observed once every two weeks, so that even though it is of a vastly different type from its neighbor SX (see next listing), observing the two together is convenient. Level 3.

160325 SX Herculis. With a period of 103 days and a range of 8.0–9.2, this star is best observed by estimating once every two weeks. Find it by moving north two degrees and then east one degree from the 5th magnitude star Pi Serpentis. Level 2.

162807 SS Herculis. Mira star, range 9.2–12.4; period 107 days. Almost one and a half degrees due south of 5.6 magnitude 28 Herculis. Because of its fast period, estimate once every week or ten days. Level 2.

173533 SZ Herculis. Range 9.9–11.8, this eclipsing binary varies over 19.6 hours. Eclipses last about 4.4 hours. In May 1963 Joseph Ashbrook of *Sky and Telescope* magazine organized a special program to observe this variable, pointing out the interesting fact that since 11 periods take 8.999 days, if you catch a minimum, you can count on another one happening precisely nine days later. The period does change minutely from time to time. Level 3.

164025 AH Herculis. A Z Camelopardalis star with a range of approximately 9–14. Look two degrees west and 0.5 degree north from 51 Herculis. Typical period is 19.8 days, with Z Cam-type standstills. Level 5.

27.5 Lyra

185243 R Lyrae. On the surface, one might consider this star an ideal one for beginners. Its range of 3.9–5.0 and its period of about 46 days seem ideal. But the chorus of problems we encounter with bright semiregulars echoes well here, especially since this star is very red. Once again, beware of the pitfalls of bright red semiregular stars with limited variation. Level 3.

191637 U Lyrae. Mira star, range 9.5–12.0; period 452 days. Level 3.

190529a V Lyrae. Mira star, range 9.7–14.8; period 374 days. Level 4.

181136 W Lyrae. Easy star to find. Simply move a little more than one degree west and then just less than one half degree north from Kappa Lyrae. W Lyrae's maximum of 7.9 is bright, and even its minimum of 12.2 can be followed in an 20 cm (8 inch) telescope. A really good star for beginners. Period 198 days. Level 2.

184134 RY Lyrae. Mira star with 9.8–14.7 range, period 327 days. The field can be found easily by moving a little more than one degree north, and then one degree west, of Beta Lyrae. The star is easy to find if it is bright, but its faint minimum is hard to see without a 30 cm (12 inch) telescope. Level 4.

183439 XY Lyrae. A bright irregular variable, easy to find at 6th magnitude. Located between Vega and Epsilon Lyrae, this star's half

Fig. 27.4. Hercules.

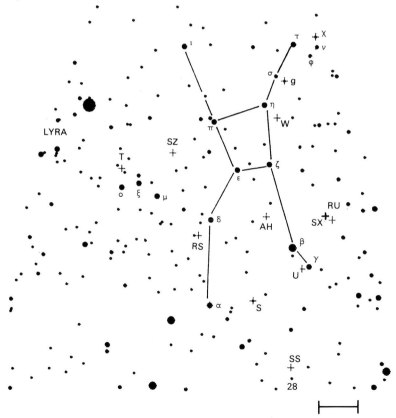

magnitude range, from approximately 5.8 to 6.4, indicates that you should not observe this star more often than once a month. Level 1.

184137 AY Lyrae. This dwarf nova star is faint, with a maximum of only 12.6 and a minimum of fainter than 16th magnitude. Situated less than half a degree from Zeta Lyrae, the field of AY Lyr should be easy to find. Level 4.

184826 CY Lyrae. This dwarf nova also is faint, with a 12th magnitude maximum, but is not difficult to find. Period roughly 17 days. Level 3.

181631 TU Lyrae. With a magnitude range of 9.3–10.3, this irregular variable can be followed throughout its range with a 15 cm (6 inch) telescope. Level 2.

Fig. 27.5. Lyra.

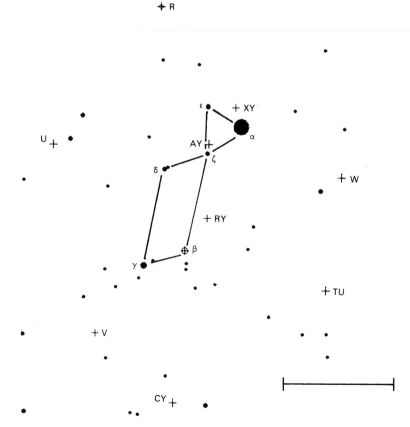

27.6 Scorpius

161122 R Scorpii. Mira star, range 10.5–14.6; period 225 days. In the globular cluster NGC 6093. Faint and difficult. Level 5.

161617 U Scorpii. A recurrent nova, maximum at 8.7, minimum around 19.3. Outbursts were observed in 1863, 1906, 1936, and 1979. Level 5.

165030 RR Scorpii. Mira star, range 5.9–11.8; period 281 days.

164844 RS Scorpii. Mira star, range 7.0–12.2; period 320 days.

173543 RU Scorpii. Mira star, range 9.0–13.0; period 371 days. Level 4.

27.7 Sagittarius

191019 R Sagittarii. Mira star, range 7.3–12.5; period 270 days. Level 2.

Fig. 27.6. Scorpius.

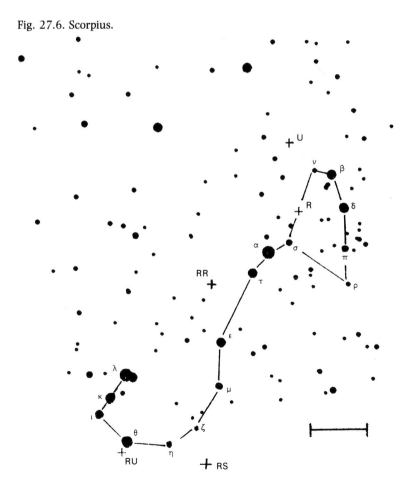

191017 T Sagittarii. Mira star, range 8.0–12.6; period 394 days. Level 2.

182619 U Sagittarii. A rare example of a Cepheid variable in an open star cluster (M25), this variable's range is 6.3 to 7.1 over 6.74 days, maximum to maximum. Writing in the July 1967 issue of *Sky and Telescope*, Joseph Ashbrook notes that there is a short standstill on the falling branch of the curve, similar to that of Eta Aquilae. Level 2.

194929 RR Sagittarii. Mira star, range 6.8–13.2; period 336 days. Level 4.

195142 RU Sagittarii. Mira star, range 7.2–12.8; period 240 days. Level 3.

190818 RX Sagittarii. Mira star, range 9.7–13.8; period 335 days. Level 4.

190819 RW Sagittarii. Semiregular variable, range 9.0–11.7; period 187 days. Level 2.

These two stars are separated by little in apparent distance and name, but by much in type of variation. RX is a traditional Mira type variable with over four magnitudes of range and 11 months of cycle. RW's period is a little over half, and its range much less, and there is a bit of uncertainty about it. You should observe them together anyway for

Fig. 27.7. Sagittarius.

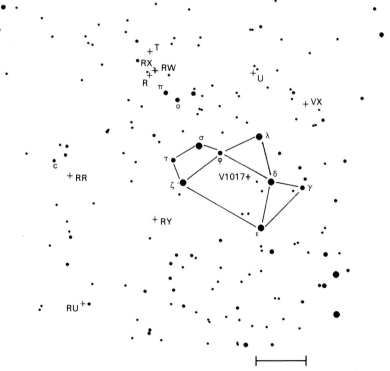

convenience; the two stars just a few minutes apart from each other, and both are near 5th magnitude 43 Sgr.

180222a VX Sagittarii. Semiregular variable, range 6.5–14.0; period 732 days. Level 3.

182529 V1017 Sagittarii. Maximum brightness 7.2, minimum fainter than 14. A recurrent nova which erupted in 1901, 1919, and 1973. Level 5.

195627 V3872 Sagittarii. Range 4.45–4.61. This bright irregular star is more commonly known as c Sgr. Not for visual estimating. Included to show that this constellation has many variables, some of which are bright!

27.8 Corona Austrina

185437 R, S, T, and TY Coronae Austrinae. These four Orion type variables tend to flicker irregularly over short periods of time. Observing this set is much more difficult than the Orion nebula family for a number of reasons. Obviously, they are harder to find, and even when you have them in the eyepiece of your telescope your troubles continue. R and TY CrA are involved in bright nebulosity, and the nebula is also variable (see chapter 22). Also, TY is near a 7.2 magnitude star. This list is included more to intrigue, since observing these stars is not recommended except for the most advanced variable star observers. Level 5.

27.9 Aquila

190108 R Aquilae. Mira star, range 6.1–11.5; period 284 days. Since R Aql happens to be in a part of its constellation devoid of bright stars, it may be hard to find. Estimate this star once every two weeks. Level 2.

195202 RR Aquilae. Range 9.0–13.9; period 395 days. Level 4.

195109 UU Aquilae. This poorly observed U Gem star varies roughly from 10.5 to 16th magnitude in a period of approximately 50 days. Level 5.

194309 OO Aquilae, a bright eclipsing variable near Altair. This

Fig. 27.8. Corona Austrinae.

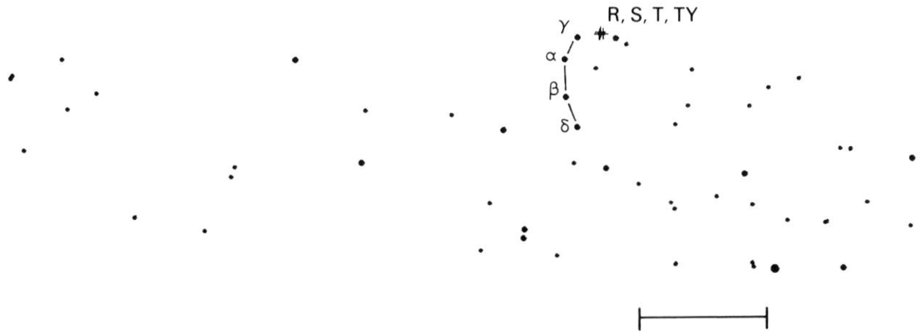

rapidly varying eclipsing binary passes through its cycle in almost precisely half a day, 12.16 hours. Its maximum of 9.2 is offset only three hours later by a 9.9 minimum. (Actually two unequal minima occur, detected with photoelectric photometers, one of 10.09 and the other of 10.05.) This variation is very slight; if you detect any change at all you are an observer either with great perception or with some imagination. Level 5.

27.10 Sagitta

200916 R Sagittae. An RV Tauri star, similar to R Scuti, except that a much longer period is superimposed on the short one. The range is

Fig. 27.9. Aquila.

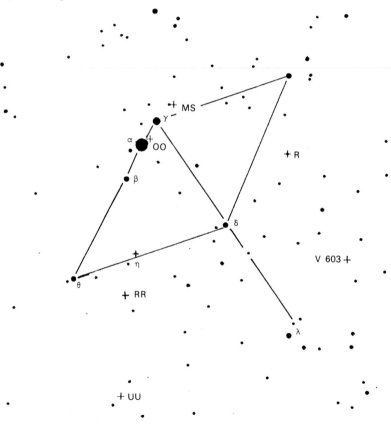

approximately 8.0–10.4. The short period is 71 days; the long one 1112 days. Estimate once a week. Level 3.

191419 U Sagittae. Eclipsing binary, range 6.5–9.3; period 3.38 days.

201520 V Sagittae. An irregular variable, this star changes in small amounts, almost nightly. V Sagittae may have been a nova a long time ago, and if this is true, it could become a nova again. It also is an eclipsing binary with a range of 0.6 magnitude. It is difficult to find; work from the 6th magnitude star 18 Sagittae and the group of stars to its east and south. Since the patterns are not too obvious, finding it may take some patience. Range 8.6–13.9. Level 4.

27.11 Vulpecula

205923a R Vulpeculae. The first discovered variable in the Fox, R is a Mira-type star with a range of four and a half magnitudes, from 8.1 to 12.6, and a period of 137 days. Find this star using 33 Vulpeculae, a 5.6 magnitude star, and a slightly fainter neighbor. Level 2.

In the same field are two other stars that could be watched from time to time. One is 205923b, VZ Vulpeculae, a star that is called a "suspected variable" although catalogs officially list it as constant at 9.6. In cases like this, we need to be careful. The star may have been given the variable designation long ago, and by mistake, based on incorrect observations. On the other hand, the observations may be quite correct, but the star simply has shown no activity in a long time. In either case, where uncertainty is present, occasional observations will not hurt.

205923c DY Vulpeculae, yet another star in the same field, has been confirmed as a variable, and is either completely irregular or does not have sufficient observational evidence to indicate periodicity. In any case, it varies by over a magnitude from roughly 8.4 to 9.7. Be careful that you always observe this red star under similar conditions of weather, sky, Moon, and telescope.

193220 U Vulpeculae. a bright Cepheid with a range from 6.7 to 7.5. The rapid rise takes much less than half the period, which is typical of this

Fig. 27.10. Sagitta.

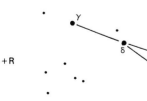

type of Cepheid. Period of 7.99 days "synchronizes" with Earth's rotation, so that if you observe a minimum at night, you should see many others at night as well. During their work to create a catalog of bright northern stars, G. Muller and P. Kempf discovered that this star was not constant.

203422a RU Vulpeculae. Semiregular variable, range 9.1–11.2; period 156 days. Level 2.

27.12 Cygnus

193408 R Cygni. Range 7.5–13.9; period 426 days. Not hard to find. Level 2.

203935 X Cygni. A bright Cepheid variable with an unusually large range of a full magnitude, from 5.9 to 6.9. The first person to notice this star's changing brightness was S. C. Chandler, in October 1886. The present period is 16.39 days, a tiny fraction longer than the 16.3855 days that the discoverer had calculated. Estimate this star once every night, especially around its sharp maximum. Level 2.

201647 U Cygni. With a moderate range of 7.2–10.7, and a very long period of 463 days, this Mira variable does not repeat its patterns from one cycle to the next. Maxima may be peaked or flat at different periods. Find it about one and three-quarter degrees northeast of the three stars that make up Omicron 1 Cygni. The *General Catalogue of Variable Stars* notes that it has been observed as high as 6.7 and as low as 11.4. You need estimate U Cygni only once a month. Level 2.

203847 V Cygni, also a Mira star with a long period, but at 421 days not as long as its neighbor U. Its range is greater, with maximum of 9.1 and minimum 12.8, a three and a half magnitude range that promises good activity if you estimate once every two weeks. Level 2.

195849 Z Cygni. Mira star, range 8.7–13.3; period 264 days. Level 4.

200938 RS Cygni. Semiregular, range 7.6–9.5; period 417 days. A bright and very red variable not far from the central star in the Northern Cross. Beautiful field of stars.

194048 RT Cygni. Period 190 days. The rise of this Mira star, from

Fig. 27.11. Vulpecula.

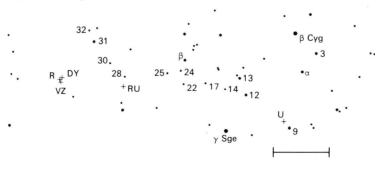

11.8 to 7.3, is steeper than its decline, with sharp minima and wide maxima. This is a good variable star if you want to catch a whole cycle in one observing season. Level 2.

213753 RU Cygni. Semiregular variable, range approximately 8.5–10.5; period 233 days. Level 3.

194029 SU Cygni. In 1897, G. Muller and P. Kempf of Germany discovered that this 6.4 magnitude star in Cygnus, drops to 7.2 in 3.84 days. This Cepheid variable is found near Beta Cygni. Level 2.

204846 RZ Cygni. This is a curious semiregular variable whose maximum varies from 9.8 to 12, its minimum from 11.8 to 14.1, according to the Kukarkin's *General Catalogue of Variable Stars*. The high and low maxima as well as the high and minima tend to alternate. The period is 276 days, but maxima can be off by as much as 0.4 of that period. Level 4.

191349 TZ Cygni. A two magnitude drop brings this 9.6 magnitude star down to 11.7 and back again in about 90 days, a relatively short time. This star is northwest of Delta Cygni, less than two degrees from the better known variable CH Cygni. Observe TZ Cygni once a week. Level 2.

Fig. 27.12. Cygnus.

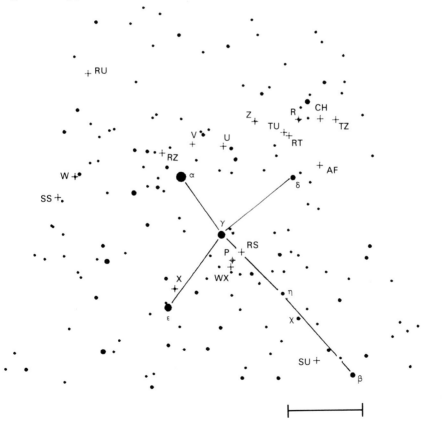

201437b WX Cygni. Less than half a degree north of the 6th magnitude star 36 Cygni, WX Cygni is a Mira variable with a range of only three magnitudes, from 9.7 to 12.6. The period is long, 410 days; thus the star that does not show much change over a short term. You need not estimate it more often than once a month. Level 3.

192745 AF Cygni. This semiregular variable star has a two magnitude range, from 6.4 to 8.4, and its period is 92.5 days. Easily found, near Delta Cygni, this star is promising for users of small telescopes. A fine variable star for a beginner. Level 1.

192545 AW Cygni. A semiregular star with a range of 8.0–10.2 and a period of about 340 days. In the field of AF Cygni. Be careful since the star is very red, a help in finding but a bane in estimating. Level 2.

Nearby is 194348 TU Cygni, with a longer period of 219 days and a range of 9.4–14.2.

193428a BG Cygni. Mira star, range 9.1–12.4; period 288 days. Level 3.

192150 CH Cygni. This is a symbiotic variable, in which a small hot white star and a large, cool red giant, orbit each other. The giant is losing gas into a shell that surrounds both stars. In some symbiotic variables, the giant undergoes Mira type variations as well.

Stars like CH Cygni are also known as combination variables because light changes can occur in several ways. The shell may vary in thickness, thus changing the total visible light from the system, the small star may vary, and the large star may vary as a Mira. Also, some stars may eclipse each other as well.

CH is a bright red semiregular variable that performs admirably for viewers with binoculars or small telescopes. The star is easily found using binoculars, and its approximate two magnitude range, from 5.6 to 8.5, is not hard to follow. Its nominal period is 97 days, but its curve is not regular. It is possible that a disk of material forms around the white dwarf only when the large, semiregular, red giant is at maximum. Look for surprises with this interesting star. Level 2.

201437a P Cygni. A so-called "nova-like" variable, that undergoes slow and very irregular variation over periods of many years. P Cygni was first seen at 3rd magnitude around 1600. P has been around 5.1 for almost 200 years. It is a very massive star, perhaps 50 times that of the Sun.

28 October, November, December

This season is variable time, with a cast of variables probably better than at any other time during the year. We still have the fine variables of the Milky Way, while toward the east, a different group of variables is gaining prominence.

This is also the time to get your fellow astronomy club members excited about the challenging field of variables. Fall is the time for renewal in many northern hemisphere astronomy clubs, where after the summer break, monthly programs and dark-of-the-Moon star parties are taking place once again. If you are fanatical about variables, you may be aware that this field of observing is not the most popular among the amateurs who attend astronomy club meetings. Observations of the changing light output of these distant suns are perceived to lack the luster of the Messier hunt or the glossy galaxy photo, and even the thrill of the meteor watch. Now is the time to insist that variables are fun.

Now we can observe Algol in all its glory, and use it as a motivation to start observing other eclipsing binaries. Two other easily found, easily observed stars are Delta Cephei, and its neighbor Mu Cephei, a huge red giant sun with totally irregular and unpredictable variations.

Another exciting star is RU Pegasi, a dwarf nova. You never know exactly when the next outburst will take place! While RU Peg may be one of the most exciting stars of fall, it surely is not the most famous. That honor is reserved for Mira itself, the type star of its class, the first discovered star of the long period variables. Mira is a queen of the fall sky, a star whose variations you can enjoy as you join company with many earlier generations of observers who enjoyed it as well.

28.1 Capricornus

201121 RT Capricorni. Range about 8.0–11.0; period 393 days. Level 3.

Fig. 28.1. Capricornus.

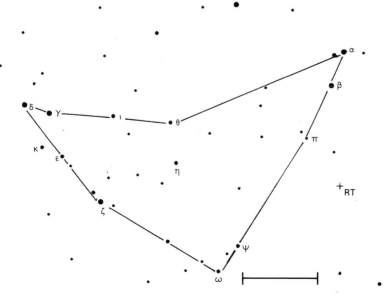

28.2 Delphinus

201008 R Delphini. Mira star, range 8.3–13.3; period 285 days. Level 2.

203816 S Delphini. Mira star, range 8.8–12.0; period 278 days. Level 2.

204017 U Delphini. Irregular variable; visual range approximately 7–8. Period about 110 days. Level 3.

28.3 Equuleus

210812 R Equulei. Mira star, range 9.3–14.5; period 261 days. Level 3. See Pegasus chart, Fig. 28.4.

Fig. 28.2. Delphinus.

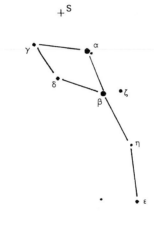

28.4 Aquarius

204405 T Aquarii. Mira star, range 7.7–13.1; period 202 days. Level 2.

204102 V Aquarii. Semiregular variable, range 7.6–9.4; period 244 days. Level 2.

223201 CY Aquarii. Range 10.4–11.6; period 0.06 days. This RR Lyrae variable (also known as a cluster variable) rises from its minimum of 11.3 to a maximum of 10.5 in slightly less than ten minutes! Although its fading is much slower, it is one of the fastest RR Lyrae variables that most northern hemisphere observers can watch. It has a period of about 88 minutes. Estimate every five minutes during falls, every one or two minutes during rises. Imagine the surprise of Alfred Jenschof of Sonneberg Observatory, when he was asked to check a suspected variable in Aquarius back in 1934, and found this! Level 3.

28.5 Pegasus

230110 R Pegasi. With a range of six and a half magnitudes (7.8–13.2), and a period of over a year (378 days), R Pegasi is a good autumn variable. It is easily found, as it forms an isosceles triangle with 55 Pegasi and 58 Pegasi, two stars of about 5th magnitude and about a degree toward the R's southeast. Level 2.

231508 S Pegasi. Mira star, range 8.0–13.0; period 319 days. Level 3.

Fig. 28.3. Aquarius.

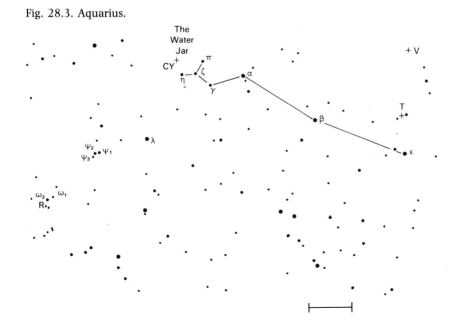

220412 T Pegasi. This Mira star varies by five and a half magnitudes, from 8.9 to 14.3, and is located over two and a half degrees north and less than half a degree east of Epsilon Pegasi. Period 379.4 days. Estimate once every two weeks. Level 4.

231425 W Pegasi. Mira star, range 8.2–12.7; period 346 days. Level 2.

235525 Z Pegasi. Mira star, range 8.4–13.2; period 335 days. Level 2.

220912 RU Pegasi. One of the best dwarf novae of the fall sky. Its 13.2 minimum is visible in a 25 cm (10 inch) amateur telescope (at least under a good dark country sky) and its maximum at 9.0 at approximately 70 day intervals is easily visible in an 15 cm (6 inch) telescope. Level 3.

222129 RV Pegasi. Mira star; range 9.9–14.6; period 397 days. Level 4.

220133b RZ Pegasi. A good Mira type variable with precisely four magnitudes of variation (8.8 to 12.8) over an unusually long 439 days. Level 3.

215927 TW Pegasi. Semiregular variable with range 7.0 to 9.2. Small variations over approximately 90 days are superimposed on a very long 956 day period. Estimate once every two weeks. Level 3.

28.6 Lacerta

222439 S Lacertae. Mira star, range 8.2–13.0; period 242 days. Level 3.

220843A RS Lacertae. Semiregular variable, range 9.6–12.5; period 238 days, Level 2.

Fig. 28.4. Pegasus.

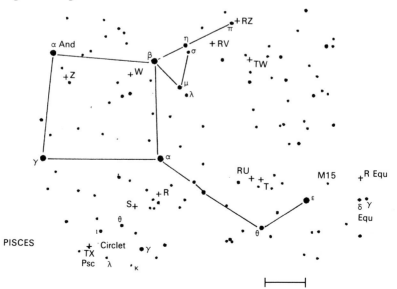

220843b RY Lacertae. Range 12.5–13.2 (photographic magnitudes). Semiregular; visual range as well as period are uncertain.

224049 RV Lacertae. Semiregular variable, range 9.5–11.4. Uncertain period, officially listed as 67 days. However, Kukarkin's *General Catalogue of Variable Stars* reports a variation in the mean magnitude over roughly two years, and the variation can apparently stop altogether for up to a year. Level 2.

225248 EW Lacertae. Tiny irregular variations of a third of a magnitude characterize this special star. In some ways it resembles an RR Lyrae star, with shallow variations over 0.7 days, but the *General Catalogue of Variable Stars* reports that not all the minima actually took place! At least this was the story for one year, 1951, but the following

Fig. 28.5. Lacerta.

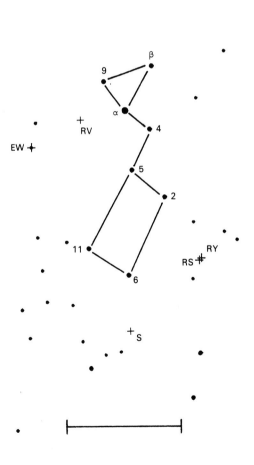

year the "period" did not follow this pattern at all. EW Lacertae is surrounded by a shell of material, and it may also have a large and variable "star spot" that may be partly responsible for these strange cycles of variation. In any case, I include this star for interest's sake, and not for actual observation, since the amplitude is too shallow for accurate visual observations.

28.7 Cepheus

213678 S Cephei. A north circumpolar star that is observable the year round from most northern hemisphere sites! This fortunate fact of position enables you to watch this star complete cycle after cycle without any breaks. It varies by six magnitudes, from 7.1 to 13.2 and back, in a period of 487 days. Enjoy a look at this red Mira star once every two or three weeks. Level 3.

210868 T Cephei. A bright Mira star two degrees south and less than two degrees west of Beta Cephei. Since it is not close to a bright star you may have to work a little to find it. Its 6.0 magnitude maximum makes it one of the brightest Miras in the sky, and even at minimum it is still at 10.3 magnitude, which allows you to follow it easily in a 15 cm (6 inch) telescope. Period 388 days. Level 1.

005381 U Cephei. An eclipsing star with a difference, U Cep is undergoing a change of period, possibly because of an exchange of gas from one star to the other. Discovered by V.K. Ceraski of Moscow in 1880, this star drops more than two full magnitudes — from 6.8 to 9.2, in

Fig. 28.6. Cepheus.

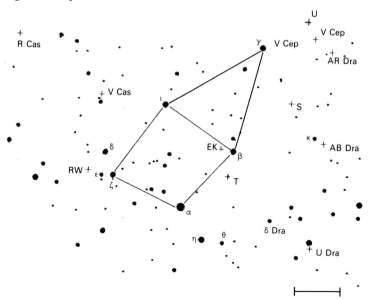

slightly less than 2.5 days. The period is changing, but in two different ways. It is increasing gradually by a very slight amount, but on rare occasions it can change more suddenly. Level 2.

213969 EK Cephei. Discovered as an Algol-type star as recently as 1959 by W. Strohmeier, a well known variable star astronomer and writer. Varies 8.2–9.5 over 4.3 days. Level 2.

Stars around the Pole

Here is a series of bright semiregular stars surrounding the pole, offering convenient objects visible at roughly the same distance from the zenith the year round to test your variable talent. Since these stars show little variation, a single observing session once a month should cover them.

225384 AR Cephei is a very red semiregular with a range of 7.0–7.9 and a roughly determined period of 116 days. Level 2.

004181 RX Cephei, near AR Cep, but with an even shorter range of 7.2–8.2 and a period of 55 days. Level 4.

235182 V Cephei is a neighbor of AR and RX Cep. It may not vary at all; I have observed it for several seasons and have observed no change. Listed as a suspected variable, it may be worth watching from time to time.

010884 RU Cephei. A red star that varies from 8.2 to 9.8 in 109 days, and 033380 SS Cephei changes from 6.7 to 7.8 over 90 days. Level 2. See Fig. 25.1.

221955 RW Cephei. Period about 346 days. Semiregular variable. Range about 8.0–10.0. Level 2.

Fig. 28.7. Some variables near Polaris.

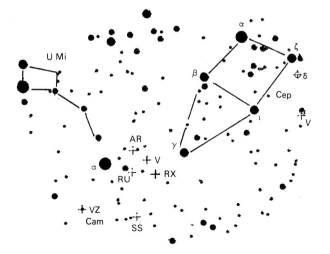

28.8 Pisces

234102 TX Piscium. A bright, irregular variable with a half magnitude variation between approximately magnitude 4.8 and 5.2. Easily found, in the semicircle of 4th magnitude stars south of the square of Pegasus. Estimate once a month. Level 1.

28.9 Andromeda

001838 R Andromedae. This Mira type star, with a period of 409 days, ranges from 6.9 to 14.3. It is easily found near Rho and Theta Andromedae, and only about five degrees southwest of Messier 31. Level 2.

010940 U Andromedae. Mira star, range 9.9–14.3; period just over 11 months, or 346 days. Difficult to find as no bright stars nearby. Level 4.

021143 W Andromedae. Range 7.4–13.7; period 396 days. This variable was discovered by a sharp observer, Rev. Thomas D. Anderson, who noticed a 9th magnitude star that had not been plotted in the famous *Bonner Durchmusterung* star atlas that became the standard for observers after its publication in 1855. In subsequent observations he noticed its fading. Level 3.

232848. Z Andromedae. Symbiotic variable that changes in complex fashion. Range 8.5–11.0. Three possible periods have been postulated,

Fig. 28.8. Andromeda.

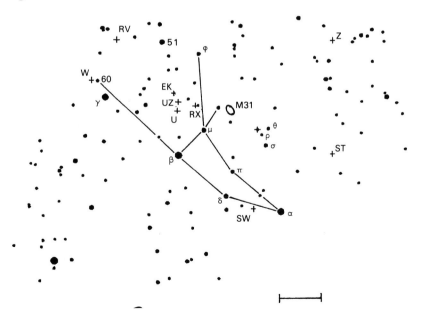

one of 300 days, another of 700 days, and a third of almost 30 years. Estimate nightly.

In this binary system, the giant is losing gas into a large shell that surrounds both stars. That shell varies in thickness, the giant undergoes Mira-type variations, and a small, hot star is varying. Level 2.

020448 RV Andromedae. Semiregular variable; range 9.0–11.5, period of 172 days. The lack of bright or well known stars nearby make this star a bit difficult to find. Level 4.

005840 RX Andromedae. Z Camelopardalis star, range 10.3–14.0; conveniently located just four degrees east of Andromeda Galaxy. Look about a third of a degree east and slightly south of 6th magnitude 39 Andromedae. Period approximately 14 days when star is not in standstill. Level 4.

233335 ST Andromedae. Mira star, range 7.7–11.8; period 334 days. Like many of the variables in the sparsely populated constellation of Andromeda, this one may be hard to find. Try moving south two and a half degrees, then east by half a degree. Estimate twice a month. Level 2.

001828 SW Andromedae. During her monumental study of the Harvard photographic plate collection, Annie Jump Cannon discovered this unusually long eclipsing binary with a period of 37 days. Its one magnitude drops, from 9.1 to 10.1, are followed by fairly flat periods around minimum that last about four hours and are followed by faster rises to maximum. The period is lengthening by about 15 minutes every ten years. Level 2.

011041a UZ Andromedae. Period 314 days, maximum of 10.1 and minimum of 14.9, so a large telescope is needed for most of this star's cycle. Level 4.

011041b EK Andromedae. Semiregular variable, range 10.3–11.4; period 185 days. Level 4.

28.10 Cassiopeia

Fig. 25.1 (p. 113) includes a chart for Cassiopeia.

005060 Gamma Cassiopeiae. In the center of the "W" of Cassiopeia is Gamma, nominally the third brightest star in the queen. This star is an irregular variable which can undergo remarkable changes over uncertain periods of time.

Although Gamma Cassiopeiae usually shines at magnitude 3.0, in 1937 it began to brighten slowly until it reached almost magnitude 1.5. Records of that remarkable event show that the star brightens slowly, over weeks, rather than quickly like a nova. Moreover, the brightening, about one and a half magnitudes, is not that unusual for a star, but when it happens to a 3rd magnitude star, that star becomes one of the brightest stars in the entire sky.

Stars like Gamma Cassiopeiae are losing mass slowly, so that the material forms either a shell that covers the star, or a disk. It may be changes in the thickness of the shell that result in the brightness changes.

In 1965, Gamma Cas again showed signs of brightening, and observers watched carefully for a repeat performance. During the observing season of 1966 I noticed some small fluctuations of a quarter magnitude over a period of weeks, but never anything more. Gamma Cas's rise is very slow so that it could sneak up on you without your being aware of any change! In fact, the appearance of the entire constellation offers a test of your visual perception. You are familiar with how the W is supposed to look; a casual glance after dinner will probably show Cassiopeia exactly as you are used to it because any small change will be absorbed by your mind and erased, so that the resulting picture is precisely what you expect.

Look at the constellation more closely. Is what you see really what is there, or has Gamma Cas shown a subtle change in brightness? Training in such perceptive skill is part of what variable star observing is about. Gamma Cas and other unusual stars help us to think about what we see, help us to see more effectively. Level 1.

235350 R Cassiopeiae. A bright maximum of 7.0 makes this an easy variable to observe *if* it is bright. Minimum at 12.6 offers a different picture, for the star is in a richly star-studded part of the sky. When R Cas is faint, you may confuse it with other stars in the field, particularly an 11.4 star very close by. Its period of 431 days is slow, but its range is high. Thus an estimate once every two weeks should be sufficient, except around maximum or minimum when you should estimate every few days. Both the maxima and the minima change considerably. Extremes of 4.8 and 8.5 have been recorded for maximum, and the star can stop at minimum anywhere between 10.4 and 13.6. Level 2.

R Cas is a beautiful sight, partly because of its unusually red color, but also because it has an 11th magnitude companion half an arc minute away. Actually this star is triple, but the 14th magnitude faintest star is very hard to see.

230759 V Cassiopeiae. Halfway between the bright stars Beta Cassiopeiae and Zeta Cephei is a Mira star called V Cas. Its closeness to 1 and 2 Cassiopeiae further eases the finding process. The star varies between 7.9 and 12.2 in 229 days. Level 2.

V Cas was discovered as variable by Thomas Anderson, of Scotland, in 1893, using a 6 cm (2.5 inch) refractor. While discoveries of new variables with instruments of this size are rare, they are not impossible.

023969 RZ Cassiopeiae. This eclipsing variable has a 1.5 magnitude drop to minimum. With so many eclipsing binaries changing by less than an magnitude, such a star would be a welcome addition to a variable star program. RZ Cas offers not only such a range, but also the advantage of being circumpolar and visible throughout the year, at least for observers in mid-northern latitudes. Moreover, the brief four hour duration of this eclipsing binary further encourages our interest. At 1.195 days, the period is unusually fast, although it changes with time, sometimes moving longer, and then becoming a bit shorter. This change is possibly caused by gravitational interaction in the system.

Variability was discovered by G. Muller at Potsdam Observatory in 1906. Range 6.2–7.7.

Around the time of an eclipse, observe this star once every quarter hour, throughout the four hour event. Level 2.

001358 TV Cassiopeiae. In a leisurely nine hour eclipse TV Cassiopeiae drops from 7.2 just over a magnitude to 8.2 and then rises again. During the rest of its 1.8 days period, TV remains at maximum. The periodicity was discovered by the famous English amateur T. Astbury, in 1911. His organized search for new variables yielded a number of new finds, including RT Aurigae and W Ursae Minoris. Level 2.

000862 UX Cassiopeiae. A Mira star with a period of 360 days, almost exactly matching our calendar year. A faint and poorly observed star with a range of about 11.5–13.8, this star is very red. Estimate once a month or even once every six weeks. Level 4.

235659 WZ Cassiopeiae. Semiregular variable, range about 9–11; period 186 days. Use binoculars and watch for the Purkinje effect in red stars, and estimate WZ Cas once a month. Level 2.

28.11 Cetus

001809 S Ceti. Mira star, range 8.2–14.2; period 320 days. Level 3.

031401 X Ceti. A Mira star with over three and a half magnitudes variation, from 8.8 to 12.3 over a period of 177 days. Only a third of a

Fig. 28.9. Cetus.

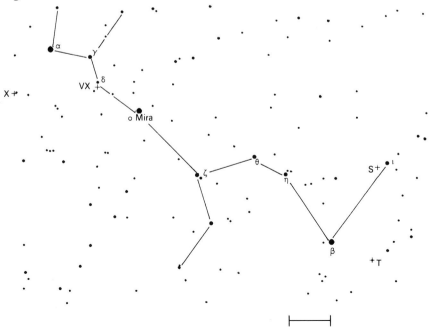

degree southeast of the 5.6 magnitude star 95 Ceti. Estimate once every two weeks. Level 2.

001620 T Ceti. Semiregular variable, range 5.0–6.9; period 159 days. Level 2.

021403b VZ Ceti. Irregular variable, range 9.5 to 12. This star is part of a binary system whose other member is Mira itself. In fact, its variation is possibly due to matter interaction with its primary, Mira. Period about 4750 days. Level 2.

28.13 Triangulum

023033 R Trianguli. Range 6.2–11.7; period 267 days. Variability discovered by T. S. Espin, an English amateur, and W. Fleming of Harvard Observatory. Estimate twice a month. Level 2.

015427 X Trianguli. Eclipsing binary that varies from 8.5 to 11.2 over a period of just 0.97 day. Variability discovered by Harvard astronomers searching through their huge plate collection in 1921. The period itself has been found to change somewhat. Level 2.

012830 Y Tri? Because this faint variable of an uncertain type varies between photographic magnitudes 14.6 and 17.2 and resides in M33.

Fig. 28.10. Triangulum.

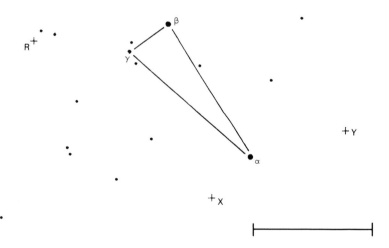

28.13 Aries

021024 R Arietis. A Mira star, range 8.2–13.2; period 187 days. Level 3.

024217 T Arietis. This semiregular variable star (range 7.5–11.3) varies over a leisurely period of 317 days. You can get to it quickly by finding 5th magnitude (Pi) Arietis, then moving a scant quarter degree west and a touch north. Do not estimate too often, however; one look a month will suffice. Level 2.

030514 U Arietis. Mira star, range 8.1–14.6; period 371 days. Level 3.

28.14 Eridanus

024312 Z Eridani. Semiregular variable, range 7.0–8.6. Primary period is 80 days; superimposed on this is another period of 746 days. Level 1.

29   Southern sky notes

For the benefit of southern hemisphere observers, I include a selection of variables that are prominent there. More information about southern stars can be obtained from the Variable Star Section of the Royal Astronomical Society of New Zealand, or the AAVSO.

29.1   Apus

145971 S Apodis. An R Coronae Borealis star with a maximum of 9.6 and minimum of 15.2. Fainter maximum, but approximately same minimum as RY Sgr. Level 4.

Fig. 28.11. Aries.

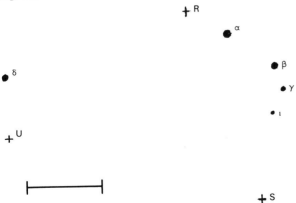

29.2 Ara

174551 U Arae. A Mira star, range 8.4–13.6; period 225 days. Level 3.

163751 V340 Arae. Delta Cephei type, range 9.6–10.7; period 20.8 days. Level 3.

29.3 Carina

104159 Eta Carinae. A marvelous variable. While in St Helena in 1677, Edmond Halley first noticed that it was as bright as magnitude 4. In 1826 it was observed at magnitude 6, but a year later it had brightened to first magnitude. By 1838 it had faded to around 1.5, but in 1843 it increased to be as bright as Sirius, the brightest star in the sky. By 1869 it was a faint 7th magnitude, and the faintest at which it has been recorded was 7.9 in 1901. Since then its variation has ranged between that and 6.5.

Fig. 28.12. Eridanus.

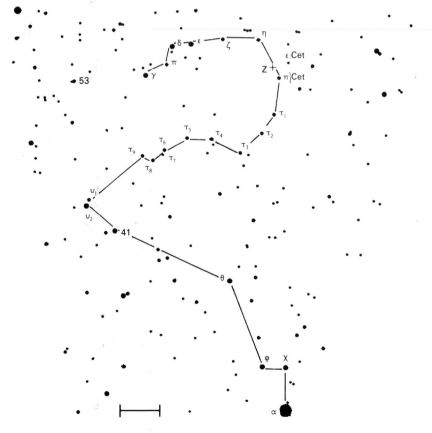

Eta Carinae, either as a single star or as a multiple object, may have close to 150 times the mass of the Sun. It is in the center of the most exquisite nebula of the Milky Way. It is one of a small number of "nova-like variables" that vary over very long periods of time. Over half a century ago, the astronomer Ernst Zinner produced a catalog of over 2000 stars based on visual magnitudes since Ptolemy, in a search for possible variations of bright stars over hundreds of years, like Eta Carinae. Even today, we classify only a few stars as nova-like.

How can anyone observe this nebula, as did Bart and Priscilla Bok, two of the century's best known astronomers, and not be moved by what they called the "grand sweep of the swirling gases?" Shortly before her death, Priscilla looked at one of their many photographs of the nebula. "After I am gone," she told her husband, "I will go to the Eta Carina nebula. That is where I want to be."

The star is embedded in so much nebulosity that it is not easy to estimate. Level 4.

092962 R Carinae. One of the brightest Miras, this variable has a range of 4.6–9.6 and a period of 309 days. Level 2.

100661 S Carinae. Another bright Mira, range 5.7–8.5; period 150 days. Level 2.

091868 RW Carinae. Range 9.3–15.0; period 319 days. Level 4.

29.4 Chamaeleon

080976 Z Chamaeleontis. U Geminorum variable. Poorly observed; maximum about 11.5, minimum 16.2; period approximately 104 days. Level 5.

29.5 Centaurus

140959 R Centauri. An unusual Mira with two maxima averaging 5.3 and 6.0 and two quite different minima of 11.8 and 8.3. Period about 546 days. Level 2.

133633 T Centauri. Semiregular variable, range 5.5–9.0; period 90 days. Level 3.

142262 V645 Centauri. This is Proxima Centauri, the nearest star to the Sun. A weak flare or UV Ceti star. Usually 13.1 magnitude, with rises to 12.1. Its variability was discovered in 1950. Level 4.

29.6 Crux

123757 AN Crucis. Eclipsing binary, range 10.6–12.3; period 3.26 days. Level 3.

29.7 Hydrus

040971 VW Hydri. U Geminorum star, range 8.4–14.4; period about 27 days. This is the most popular dwarf nova for observers in the southern hemisphere. The magnitudes listed were obtained from photographic plates; thus the visual magnitudes are different. Level 2.

29.8 Microscopium

203329 R Microscopii. Mira star, range 9.2–13.4; period 139 days. Level 4.

212030 S Microscopii. Mira star, range 9.0–13.8; period 210 days. Level 4.

29.9 Musca

123568 R Muscae. Delta Cephei star, range 5.9–6.8; period 7.51 days. Level 3.

120769 S Muscae. Delta Cephei star, range 5.9–6.5; period 9.66 days. Level 3.

29.10 Norma

152849 R Normae. Mira star with behavior resembling R Cen. Two maxima are 7.2 and 8.2, the minima are 13.2 and 9.7. Period 507 days. Level 3.

153654 T Normae. Mira star with range 7.4–13.2; period 241 days. Level 3.

29.11 Octans

055686 R Octantis. Mira star, range 7.9–12.4; period 405 days. Although its period is unusually long, this star is easy to follow since it is fairly bright and circumpolar (always above the horizon) from almost all southern latitudes. Level 2.

172486 S Octantis. Mira star, range 8.4–13.5; period 259 days. Circumpolar. Level 3.

205782 T Octantis. Mira star, range 9.8–14.0; period 218 days. Circumpolar. Level 4.

131283 U Octantis. Mira star, range 7.9–13.6; period 308 days. Circumpolar. Level 3.

29.12 Pavo

180363 R Pavonis. Mira star, range 8.5–13.0; period 229 days. Level 3.

194659 S Pavonis. Semiregular variable, range 6.6–10.4; period 381 days. Level 2.

29.13 Pyxis

090031 T Pyxidis. Recurrent nova, maximum 7.0, minimum 15.8. Outbursts observed in 1890, 1902, 1920, 1944, and 1966. Level 5.

29.14 Telescopium

200747 R Telescopii. Mira star, range 8.6–14.8; period 462 days. Level 4.

29.15 Tucana

223362 T Tucanae. Mira star, range 8.1–13.2; period 250 days. Level 4.

4

A miscellany

30 Stars and people

Variable star observing was something people had to appreciate and discover. As an observing discipline, it was forced on us time and again as bright exploding suns intruded in our otherwise placid sky. Opening and closing suddenly with each major nova, observers could watch these intruders with fear and with questions, but these stars did not appear to have any lasting effect on early civilization. Novae and comets shared the uncertain interpretation of being signs of something else, rather than being objects of intrinsic interest.

Ancient Chinese and native American records of supernovae are the earliest surviving variable star reports. The "guest star" of 1054, for example, appeared at a time when these people would be looking beyond, to new things, and would keep careful records of anything extraordinary.

The history of variable star observing closely parallels the sudden launching of Renaissance astronomical curiosity. The quiet, orderly sky of Ptolemy offered comfort with safe spheres, all fixed in content, of which the stellar was the most distant or purest.

In 1572 a bright new light punched a hole through that sphere, shattering the concept of order without change. Observations by Tycho recorded not only the appearance of this incredible sun but also its changing brightness over time. The sphere had changed, but the thing that changed it was changing in itself! If the conservative scientific establishment wanted to believe that the supernova in Cassiopeia was an aberration not to be concerned with, a second new star in 1604 added to the growing evidence that the sky was a dynamic picture. By this time Tycho was gone, and Johannes Kepler made careful notes on the changing brightness of this star.

The supernova of 1604, as it was later to be called, attracted wide public attention, both for its strange appearance and for the shaken belief in the system it dared to challenge. Shakespeare had already made passing, if facetious, reference to the controversy of "new stars" in this letter from Hamlet to Ophelia:

> Doubt that the stars are fire;
> Doubt that the sun doth move;

Doubt truth to be a liar;
But never doubt I love.

O dear Ophelia, I am ill at these numbers; I have not art to reckon
my groans: but that I love thee best, O most best, believe it. Adieu.

30.1 John Goodricke

A faint comet in Cygnus finally gave away Algol's secret.

Our story begins in 1764, when a boy newly born in Groningen,
Holland, was found to be totally deaf. The son of enlightened parents who
did not share the contemporary belief that deafness and stupidity were
equated, John Goodricke was sent to a special school in Edinburgh, a
school whose headmaster had developed a process to teach deaf children
to speak and to think. Meantime, his parents moved to York, England, by
fortunate accident just a few houses from the Pigott family, in which
Nathaniel and his son Edward were astronomers. When John Goodricke
returned to live with his parents, he had developed at least a theoretical
interest in astronomy, and quickly became friends with Edward. With
Herschel's discovery of Uranus taking place at the same time, the two
young men must have had much to consider.

On November 15, 1781, Edward independently discovered a new
comet "with a small nucleus & coma near the neck of Cygnus." This was
actually Comet Mechain, 1781 II, which had been discovered five weeks
earlier by the French observer whose name it bears. By the time Pigott
observed this comet, it was a conspicuous object of at least 4th magnitude,
with a tail four degrees (eight full moons) long. Seeing this comet inspired
Goodricke to begin a diary of personal astronomical observations, and he
described his friend's discovery on its opening page. When Edward
suggested a few months later that he begin a search for new variable stars,
John responded immediately and with enthusiasm. On November 12,
almost a year after the Cygnus comet, John observed a sudden drop in
Algol's brightness. Edward was certain that Algol was variable, but
neither had any idea that the change would come so quickly. Six weeks
later, on December 28, the two friends observed the variation again.

In 1667, Geminiano Montanari, a sharp observer from Bologna, Italy,
had noticed an occasional drop in the brightness of Beta Persei. Although
his observations clearly recorded a sharp change in brightness, something
not observed before except for novae, his fellow astronomers did not
continue studying this star, and it was not until 1782 that Goodricke and
Pigott determined the incredibly short period of about 69 hours. For Algol,
Goodricke offered two possibilities. One was that the star was partly
covered by large dark markings or spots that would, through rotation,
cause the drop in brightness. The other theory was that a companion
object revolved around the star. Not until more than a century later,
during the 1880s, did theoretical work by E. C. Pickering and observa-

tions with a spectroscope by H. C. Vogel finally give voice and hearing to the work of the 18 year old Goodricke.

The night of September 10, 1784, did much for variable star astronomy, for the two friends independently discovered two of our best known variable stars. As Edward was detecting variability in Eta Aquilae, John observed a brightness change in Beta Lyrae. A month later, Delta Cephei revealed its secret to Goodricke's astute eye.

What a successful career for someone so young! By the age of 19 Goodricke had discovered the variations of three stars. But his life would not permit further joy and further discovery. In early April of 1786, after a series of Delta Cephei observations on what must have been cold nights, he became ill and died on April 20.

Edward was shattered, losing his joy in observing. He had been away from York at the time, and simply continued his wanderings for over a decade. Fortunately he did recover his drive to be among the stars, and during his travels he discovered that two more stars vary, stars that today are still of great interest, R Scuti and R Coronae Borealis. As he got older his enthusiasm drained. Feeling that he had not been credited for his variable star work, he apparently gave up astronomy around 1810.

Carolyn Gilman's article, "John Goodricke and his variable stars" (*Sky and Telescope*, November 1978, pp. 400–3) develops further the story of what Nathaniel Pigott happily called "the three York astronomers." Of the three, John Goodricke is the best remembered. His genius in discovering the period and correctly interpreting the cause of Algol's variation may have been aided by the silent world he was forced to endure, relying more heavily on his sight and his mind to see his world and to interpret his observations.

30.2 AAVSO, observing as community

On the evening of July 6, 1936, William Tyler Olcott was introducing astronomy to a small group of summer residents at George's Mills, New Hampshire. Olcott loved the stars, but especially the variables because he knew that in their pulsations he was seeing astronomy evolve; he was watching the Universe happen. He enjoyed lecturing about them, writing poems about them, and watching with pride the steady growth of the American Association of Variable Star Observers which he had founded twenty-five years earlier.

The audience listened carefully as the lecturer shared his commitment to astronomy, first with excitement, later with some concern as the speaker began to grow weak. In the middle of his talk he began to grasp for breath and finally he collapsed from a heart attack. His subsequent death stunned his friends and the members of his beloved AAVSO.

Many of these early members had discovered the beauty of the variables through Olcott's words. At the end of his *Field Book of the Stars*, published early this century, he had written:

"Many readers of this book may be the fortunate possessors of small telescopes. It may be that they have observed the heavens from time to time in a desultory way and have no notion that valuable and practical scientific research work can be accomplished with a small glass. If those who are willing to aid in the great work of astrophysical research will communicate with the author he will be pleased to outline for them a most practical and fascinating line of observational work which will enable them to share in the advance of our knowledge respecting the stars. It is work that involves no mathematics and its details are easily mastered."

Imagine the numbers of people, especially young people, reading this invitation and rushing to join in the work of variable stars! Remember that this was the time of the opening of the Universe with the opening of the 100 inch (2.5 m) Hooker telescope on Mt Wilson. Astronomy was on the move, and here was a chance for amateurs to help it along.

With the help of his early followers, Olcott transformed the AAVSO into a respected observational organization. Today it thrives as a living reminder of Olcott's most ardent hope, that amateur and professional could work closely in astronomy. Olcott would have been flabbergasted to know of the High Energy Astrophysical Observatory (HEAO) satellites, as well as the International Ultraviolet Explorer (IUE) that have orbited the Earth with programs to observe stars in their outburst stage. Announcements of these outbursts came in part from observations of AAVSO amateurs. Olcott felt that the AAVSO would bring amateurs and professionals together, and benefit three groups, the amateur, who loved to follow the behavior of a favorite variable star, the professional observational astronomer, who needed to know what that star was doing to plan precious observing moments, and the theoretical astrophysicist who needed the observations to understand the star.

The AAVSO is an organization of extraordinary people who have found in variable stars a special motivation to pursue their astronomy. These are people who began with small telescopes, looking at the Moon, planets and double stars. A minority of the AAVSOers have found variables so satisfying that they observe little else, but most of today's AAVSO members are also avid observers of galaxies, planets, comets, and everything else the sky has to offer.

I became entranced by the magic of variable stars two decades ago through reading *Starlight Nights*, Leslie Peltier's stirring autobiography. Today I am still addicted to the magic of what the AAVSO stands for. Variables offer a special way to commune with the stars, a unique opportunity to deal with each star in your program on its terms.

The AAVSO is more than that. In addition to its main purpose of gathering and reducing amateur data, it also offers programs in nova search and in solar observation. Its members gather twice a year to share their observing experiences. In the spring of 1980, the AAVSO met during a raging thunderstorm in Houston. A year later, its members gathered among the domes and peaks of Tucson. Each fall, the group meets somewhere in Massachusetts, often in the Boston area, but sometimes out at more exotic places like the Maria Mitchell Observatory on Nantucket Island.

At the helm of the organization is its Director, for many years Margaret Mayall, and since 1973, Janet Mattei. Overseeing the thousands of observations that pour in each year, organizing the two annual meetings, and answering correspondence with members, interested schoolchildren, and professional astronomers, requires a good deal of energy and enthusiasm. Each new AAVSO observer receives a package of 10 introductory charts. For over 50 years the set has included maps of R Leonis, a long period variable not far from Regulus. Experienced and novice observers alike watch this Mira star, and among the observers who began with it was Leslie Peltier of Ohio, whose single look at R Leonis began a string of 132000 variable star observations.

30.3 Leslie C. Peltier

If the enthusiasm behind amateur variable star observing during the last half of the twentieth century could be traced to one man, it would be Leslie C. Peltier of Delphos, Ohio, a man who taught us much about what it really means to be an astronomer.

Born on January 2, 1900, Leslie C. Peltier grew up on a farm near Delphos, Ohio. His interest in astronomy was sired by sightings of two major comets in 1910, but the real spark came on a May evening in 1915 when he looked up, possibly at a meteor, and wondered why he did not know any of the stars above him.

Peltier picked 900 quarts of strawberries on the family farm, at 2 cents per quart, to raise the money that bought his first telescope. Those of us who know and love R Leonis (see chapter 11) have a hard time thinking about it without also thinking of Peltier's special relation to this curious star. He began his observations using a 2 inch (5 cm) brass telescope mounted on a heavy stand with a discarded grindstone as its azimuth support.

Leslie's first night out was not auspicious. "As soon as darkness fell," Leslie writes in his autobiography, *Starlight Nights*, "I bundled up and, with telescope, atlas, and charts in my mittened hands, I went out to find my variables. Two hours later, when I returned, half frozen, to the fire I had not found a single one."

After several more nights of trying, Leslie finally found R Leonis. He attributed his problem to not knowing how much of the chart his telescope was showing in a single field of view. Once learned, his lesson was not forgotten, and as Peltier's reputation as a skilled observer spread, he was offered the use of a short focus 15 cm (6 inch) refractor by Henry Norris Russell of the Princeton Observatory. Peltier learned that this telescope had been used in the discovery of three comets by Zaccheus

Daniel, including an impressive naked-eye object, in 1907. With that telescope Peltier would independently discover 12 more comets.

The first time I visited Leslie Peltier was near the end of a cross-country journey to visit some major professional observatories in the spring of 1974. Leslie's site was every bit as imposing as the giants of Palomar, and in a sense, even more so, because I was about to meet the director as well, a man who had discovered comets and had completed thousands of variable star observations. The beautiful front lawn, punctuated by an enormous white oak, beckoned a visitor to enter. I walked into a huge living room, decorated with some fine antique furniture. But the feeling of another era ended right there. In a conversation about comets, variable stars, and politics, Leslie was as vibrant as he had been in his writing. His last comet had been found in 1954, and since then he had devoted much of his observing life to variable stars. "I don't think amateurs have it so easy any more in comet hunting," he said. "Professional photographic searches make it quite difficult for an amateur to come in first." Should we give up then? "Of course not. You just have to try harder."

For Leslie Peltier, astronomy was more than a hobby and his work with variables more than an interest. He had a sense of purpose in the observations he made that kept his family of variables close to his consciousness every waking moment. More than the numbers, Peltier had the secret ingredient of perseverance that came from his heart, a sense of commitment both to the variables he loved and to the spirit of scientific inquiry. He was primarily interested in how his variable stars behaved, and the large numbers of estimates naturally followed. Be careful, when you observe, not to reverse these priorities.

In variable star observing, both his work and his methods are still used by the AAVSO and by the professional astronomers who use the data. With so many variable star observations, comet and nova discoveries, articles, and astronomical books, his reputation is solid. But he was very modest about this record, and acted as though it did not mean that much to him. At the close of my final visit with Peltier, in 1979, I told him that a magazine was interested in interviewing him. He shook his head thoughtfully. "If they really want to, that's fine," he replied. "But I would think they would want a professional astronomer, not just an amateur like me."

31 The next generation

Through the magic of variable stars you now have passed through a looking glass. You have discovered stars passing through the eruptions and inconsistencies of youth and you have visited and drawn comfort from the cosmic wisdom of the aged Mira stars. You have watched the dwarf novae perform and you have looked on as old novae sleep peacefully as the eons after their outbursts pass by.

You passed through the looking glass in a mood of curiosity and I hope you return unsatisfied, demanding more. Through the bibliography, I have suggested certain sources that may help, but the most important source of all is that of the glass of your telescope coupled to the retina of your eye and connected by optic nerve to an inquiring mind searching for answers. Every year new variables are discovered and much more is learned about how these stars behave.

Quite possibly your own observations may add to this knowledge. Possibly also, you will want to pass this on to someone younger.

In the nature of variable stars lies a key to introducing children to astronomy. Stars are individual, they behave in definite ways. They are born, they grow up, they mature, they grow old, and they die. Stars also have moods — sometimes their light output changes. When a star is young it might flicker with the intensity of a rebellious child, unable to decide on its future course. As it matures it acquires the dignity of middle age, staying out of trouble and perhaps taking the responsibility for lighting a family of planets.

Old age does strange things to a star. Now a red giant, it may begin varying again, but this time to the slow and measured beat of an aging heart. Or the star may become unpredictable, varying over time with graceful but completely irregular pulses. A more massive star might vary as it struggles for its last gasps of hydrogen: tearing itself apart in a final cry for help, it will stun the galaxy as it bows out of active life — and even then its tiny heart may survive to vary some more as the clock-like tick-tock of a pulsar.

In teaching children the sky, variable stars are a key to understanding a sky that performs. If we amateur astronomers can aid an understanding of science by observing variable stars, we can make a second important contribution by sharing our love and interest in the stars with a child. We are in a unique position to do this; our interest began not through the textbook but through the stars directly. Wherever children gather, with families, classrooms, or in summer camps, we can show them how to discover the variables. Not just high school students can benefit; younger grade school children, aged 7 to 12, could also discover astronomy through the magic of variable stars.

The most important thing is not to transfer information to the next generation, but to convey your love of the subject. Emphasize that variable stars don't just shine in the sky, they perform. The young Orion variables enjoy the unpredictability of childhood, the U Gems have temper tantrums, the huge Miras vary with the majesty of senior citizens.

Anthropomorphizing the variables is intended as an aid to inspire. It is a useful teaching tool so long as the children do not go home believing that stars are really just like us. People are people and stars are stars, and if some apparent behavioral similarities can open a gate to their interest, then we and the children can enjoy them.

In observing an active, changing star, a child discovers it, and thinks of questions. What would life be like on a planet orbiting such a star? Was the star always like this? Will our Sun someday be like it? Young children can observe variables if the stars perform regularly and if they and their comparison stars are easy to find. Delta Cephei is my favorite for young people since it satisfies these conditions so well. With a period of 5.37 days, its full cycle possibly can be seen and understood in a typical school week. The two comparison stars form an easily observed triangle with it. The star will show some change with each observing night, so it should keep the level of interest high.

Algol is interesting partly because it is so famous. It is bright, and by simply choosing the next convenient minimum, a child can watch it perform. Since Algol's minima do not often happen early in the evening, a careful teacher can schedule a special observing party when it does. Betelgeuse is a good star for different reasons. It is so bright and red, and part of a major constellation, that finding it should present no problem. Its strange name also would attract a child's attention. Although estimating Betelgeuse is fairly difficult since the star is red, its change is irregular, and comparison stars are far across the sky, it is a good representative of its type.

Of all the Mira stars, two seem appropriate for children, Mira itself and Chi Cygni. With both stars, planning an observing program in advance is important so that the children do not look for them when they are faint. Mira is quite easy to find when it is brighter than 5th magnitude, and from a dark site, Chi Cygni changes the appearance of the long arm of the Northern Cross (Cygnus) when it is near maximum.

Understanding the motivations of people who devoted their lives to this aspect of science is just as important, and just as inspiring, as observing their stars. The life of John Goodricke is especially interesting since his story includes rising from the handicap of being unable to hear or speak. His discovery of the variation of Algol was one of the most significant events in the advance of astronomy, and it happened from a back yard. Through his story, the variable stars and those who study them can come closer to children.

Variable stars teach us that the starry sky is a different place, that it is accessible to young people, and that by following these stars we become a part of them. It is hard to end an observing session with variables without feeling changed, without feeling that you have reached through your telescope and touched a star.

32

Going further

32.1 Books

Because this book has been designed to interest you in the observation of variable stars, it is not filled with graphs and formulae or explanations of variation. There are other books that tell you that, as well as other branches and ideas of astronomy, and here are some of them.

Bishop, Roy, ed. *The Observers Handbook*: annual. Toronto: The Royal Astronomical Society of Canada. One of the best of the annual guides to the night sky.

Calder, Nigel. *Violent Universe.* New York: Viking, 1969. This vibrant book injects a healthy dose of enthusiasm into what is known about the expanding Universe.

Campbell, Leon. *Studies of Long Period Variables.* Cambridge, Mass.: AAVSO. This 247 page volume, long out of print, is a landmark study of almost 400 long period variables based on decades of AAVSO observations.

Clark, David H., and Stephenson, F. Richard. *The Historical Supernovae.* Oxford, Pergamon Press, 1977. A useful discussion of ancient supernovae.

Dickinson, Terence. *The Edmund Mag Six Star Atlas.* Barrington: Edmund Scientific Company, 1982. A good atlas by a man who loves the stars.

Dickinson, Terence. *NightWatch.* Camden East, Ontario: Camden House, 1983. A superb introduction to "viewing the Universe." Chapters on variables and other branches of observational astronomy.

Edberg, Stephen J., and Levy, David H. *Observe Comets.* Washington: Astronomical League, 1985.

Harwit, Martin. *Cosmic Discovery.* New York: Basic Books, 1981. Discussion of the process of discovery in astronomy.

Hoffmeister, C., Richter, G., and Wenzel, W. *Variable Stars.* trans. Dunlop, S. Berlin: Springer-Verlag, 1985. A professional introduction to the many forms of variation in stars.

Hogg, Helen Sawyer. *The Stars Belong to Everyone.* Toronto: Doubleday, 1976. This Canadian professional astronomer has written a superb introduction to astronomy. It is a friendly book; sit down by the fireplace and enjoy looking through its pages.

Howard, Neale E. *The Telescope Handbook and Star Atlas.* 2nd ed. New York: Crowell, 1975. Some very clear diagrams and good instruction regarding the use of a telescope.

Kukarkin, B. V., *et al. General Catalogue of Variable Stars.* Moscow: Astronomical Council of the Academy of Sciences in the USSR, 3rd ed., 1969. 4th ed., 1985. The three volumes and three supplements of the 3rd edition of this valuable research source list every known variable at the time of its publication. The lists of variable types at the beginning of volume one is a noteworthy summary.

Levy, David H. *The Joy of Gazing.* 2nd ed. Montreal: Montreal Center of the Royal Astronomical Society of Canada, 1985. A small book that presents ways of getting started in several areas of observing.

Levy, David H. *The Universe for Children.* Oakland, California: Everything in the Universe, 1985. How astronomy-minded adults can help children to love the sky.

Levy, David H. "Variable Stars." This column for amateur astronomers appeared in several issues of *Star and Sky* magazine in 1980 and early 1981, in *Deep Sky Monthly* magazine until 1982, and continues in *Deep Sky* magazine.

Levy, David H. and Edberg, Stephen J. *Observe Meteors.* Washington: Astronomical League, 1986.

Menzel, Donald H. and Pasachoff, Jay M. *A Field Guide to the Stars and Planets.* 2nd ed. Boston; Houghton Mifflin, 1964. Part of the Peterson Field Guide Series. Compact and thorough.

Mitton, Jacqueline. *Astronomy: An Introduction for the Amateur Astronomer.* New York: Scribner's, 1978. A compact, well-written introduction with good explanation of the Hertzsprung-Russell diagram of the evolution of stars.

Mitton, Simon. *The Crab Nebula.* New York: Scribner's. The story of the supernova of 1054 and the nebula and pulsar that remain.

Muirden, James. *The Amateur Astronomers Handbook: Revised Edition.* New York: Crowell, 1974. A good basic compendium with a chapter on variable stars.

Newton, Jack, and Teece, Philip. *The Cambridge Deep-Sky Album.* Cambridge: Cambridge University Press, 1983. A collection of amateur photographs of Messier and other objects. The galaxy photos are useful sources for searchers of supernovae.

Norton, Arthur P. and Inglis, J. Gall. *Norton's Star Atlas.* ed. Gilbert Satterthwaite, 17th Ed. Cambridge, Mass.: Sky Publishing, 1978. Although the text is at times esoteric, this is one of the best portable star atlases.

Pasachoff, Jay M. and Kutner, Marc L. *University Astronomy.* Philadelphia: W. B. Saunders, 1978. Thorough introductory textbook.

Paul, Henry E. *Binoculars and All Purpose Telescopes.* New York: Amphoto, 1964. Good handbook for prospective purchasers.

Paul, Henry E. *Telescopes For Skygazing.* 3rd Ed. Garden City: Amphoto, 1976. Written by a man who loved telescopes, this book is a must for telescope owners. Easy to read, many illustrations. The Henry Paul Books can also be obtained from Sky Publishing Corp.

Peltier, Leslie C. *Starlight Nights: The Adventures of a Star Gazer.* Sky Publishing, 1980. A captivating, exquisite study of the life of a man who observed the night sky for over 60 years. No other book captures so well the emotion of an amateur astronomer who loves the stars. No other book answers more powerfully the question of why people go out at night to watch the stars.

Peltier, Leslie C. *Guideposts to the Stars.* New York: Macmillan, 1972. Leslie Peltier's second book reaches out to beginning star lovers. I highly recommend this book to all who would like to have fun as they learn the sky and the constellations.

Peltier, Leslie C. *Leslie Peltier's Guide to the Stars.* Milwaukee: AstroMedia, and Cambridge: Cambridge University Press, 1986. A beautifully written guide to enjoying the sky with binoculars.

Sagan, Carl. *The Cosmic Connection.* New York: Dell, 1973. Penetrating comments on the importance of space. Well written and lively.

Sagan, Carl. *Cosmos.* New York: Random House, 1980. This book as well as the television series upon which it is based give you a feeling for the beauty and the romance of astronomy.

Salmi, Juhani. *Check a Possible Supernova.* Vesijarvenk, 36 C 40, 15140 Lahti, Finland. 1984, 1985. This astrophotographer has published two good sets of galaxy photographs, 40 cards each. Spiral bound, they provide a convenient reference source for supernova hunters.

Scovil, Charles E. *The AAVSO Variable Star Atlas.* Cambridge, Mass.: Sky Publishing Corporation, 1980. Excellent guide for finding variable stars. Also useful for finding and estimating comets and asteroids.

Sidgwick, J. B. *Introducing Astronomy.* Rev. ed. London: Faber, 1973. This book is especially useful for its appendix in which each constellation is drawn and described along with the exciting things to see, including bright variable stars. I started with this book, maybe you should too!

Sidgwick, J. B. *Observational Astronomy for Amateurs.* 4th Ed. Rev. Muirden, James. Hillside, N. J.: Enslow Publishers, 1982. Excellent section on methods for observing variable stars.

Strohmeier, W. *Variable Stars.* New York: Pergamon Press, 1972. A good graduate-level introduction to variable stars.

Tirion, Wil. *Sky Atlas 2000.0* Cambridge, Mass.: Sky Publishing Corp., and Cambridge, England: Cambridge University Press, 1981. Very good atlas of the entire sky to magnitude 8.0. Most variable stars that reach magnitude 8.0 are also plotted.

Thompson, Gregg and Bryan, James. (Searching for Supernovae). Cambridge: Cambridge University Press, 1987.

Magazines

Astronomy, 1027 N. Seventh St, Milwaukee, WI 53233. A monthly magazine with excellent articles both observational and theoretical astronomy. Perfect for adult beginners.

Deep Sky, 1027 N. Seventh St, Milwaukee, WI 53233. Quarterly magazine with emphasis on observing objects outside the solar system. Includes a column on variable stars by the author.

Odyssey, 1027 N. Seventh St, Milwaukee, WI 53233. Monthly magazine meant for children. Introduces the sky gracefully to the next generation.

Sky and Telescope, 49 Bay State Road, Cambridge, MA 02238. A fine monthly magazine with regular features devoted to observing, including a list of variable stars approaching maximum.

Star Date, RLM 15.308, The University of Texas at Austin, Austin, TX 78712. Excellent beginner-aimed monthly "news report" and observing guide.

32.2 Some major amateur variable star organizations

United States and Canada American Association of Variable Star Observers, 25 Birch Street, Cambridge, MA 02138, USA

Great Britain British Astronomical Association Variable Star Section, Douglas R. B. Saw, Upanova, 18 Dollicott, Haddenham, Aylesbury Bucks, HP17 8JG, England

New Zealand Variable Star Section of the Royal Astronomical Society of New Zealand, Variable Star Section, Dr. Frank M Bateson, Director, PO Box 3093, Greerton Tauranga, New Zealand

France Asociation Francaise d'Observateurs d'Etoiles Variables, Emile Schweitzer, 16, rue de Plobsheim, 67100 Strasbourg, France

33 Glossary and abbreviations

33.1 Glossary

Absolute magnitude: The brightness a star would appear to have if it were about 33 light years, or 195 x 10 to power of 12, miles, from us.

Amplitude: The range in brightness from maximum to minimum.

Apparent magnitude: A star's brightness as seen from Earth.

Binary system: A system in which two stars revolve around each other.

CCD: A charge-coupled device that uses a light-sensitive electronic chip to record images through a telescope.

Cepheid: A supergiant star that varies in brightness with extreme regularity; periods from 1 to 60 days.

Cluster variable: Cepheid variable with period of one day or shorter. Found in globular clusters.

Degree: A unit of measurement in the sky, approximately equal to two full Moons.

Dwarf nova: A star with periodic outbursts, months apart, during which its brightness rises by three to four magnitudes. "Dwarf" refers to the size of the outburst, not to the presence of a white dwarf in the system.

Eruptive star: A star in which an explosion or an eruption plays a part in the process of variation.

Eclipsing binary: A double star system in which one member periodically eclipses the other, so that we observe a variation in brightness.

Field of view: A measure, usually in degrees, of the amount of sky visible in binoculars or telescope.

Globular cluster: A spherical grouping of (usually) old stars. Occurs most often above or below the area of a galaxy's stars.

Hump: Occasionally a fading variable will brighten for a short time, or a brightening star will fade briefly. When such behavior is plotted on the star's light curve, it is known as a hump.

Intrinsic variation: Light changes in a star that arise from its fundamental nature.

Ionize: To convert into ions, atoms that have gained or lost one or more electrons.

Julian day: To ease the keeping of records that span months and years, variable star observers use this consecutive count of the number of days since noon on January 1, 4713 B.C.

Light curve: A demonstration of a star's variation, in which observed magnitudes are plotted against time.

Light year: The distance light travels in a year.

Long period variable: A star whose brightness oscillates regularly over a period of months. Known also as Mira star.

Magnitude: A measure of the brightness of a star.

Messier, Charles: French comet hunter who began compiling a catalog of objects to avoid, in 1758. His lists of star clusters, nebulae, and galaxies were published in 1771 (45 objects), 1780 (68 objects), and 1784 (103 objects). Pierre Mechain, also a comet hunter, added six more objects. Catalog objects are listed in the form Messier 31, or M31.

Mira star: A long period variable star whose brightness oscillates regularly over a period of months.

Nebula: A cloud of gas and/or dust around a star or in interstellar space.

NGC: The *New General Catalog*, of almost 8000 nonstellar objects, like star clusters, nebulae, and galaxies, originally compiled by Johan Ludwig Emil Dreyer for publication in 1888.

Nova: A star system that brightens rapidly and unpredictably by several magnitudes.

Open star cluster: A grouping of relatively young stars, much smaller than a globular cluster.

Orion variable: Young star with irregular and often rapid variations.

Photoelectric photometer: A device used to measure light from an object.

Period: The length of time between two maxima or two minima of a variable star.

Precession: A slow wobbling of a body, like the Earth, on its axis.

Proper motion: The actual motion of a star across the sky as it moves through space.

Range: The amount of brightness change in a variable star.

Red giant: A large, cool star, usually 10 to 100 times the Sun's diameter.

R Coronae Borealis star: A variable whose long periods of constant maximum brightness are interrupted by sudden drops of several magnitudes, followed by slow rises.

Recurrent nova: A nova for which more than one outburst has been observed.

Semiregular variable: Large red stars whose periods and amplitudes of variation are not as regular as those of the Mira stars.

Standstill: A period of constant brightness in a variable star.

Supergiant: An extremely luminous star, 10 to 1000 times the Sun's diameter.

Supernova: A star that brightens by many magnitudes to rival the brightness of an entire galaxy.

Symbiotic variable: A star system with irregular and complicated variation that sometimes includes long term brightenings.

T association: A grouping of Orion type stars that have formed from a cloud of dust and gas.

UV Ceti star: Irregular variable showing strong outbursts of several magnitudes taking place on a time scale of minutes. Also called flare star.

Variable star: A star that changes in brightness.

White dwarf: A star the size of a planet, approximately the Sun's mass.

33.2 The Greek alphabet

α	Alpha	(*Al*-fah)	ν	Nu	(*Noo*)
β	Beta	(*Bay*-tah)	ξ	Xi	(*Z-eye*)
γ	Gamma	(*Gah*-mah)	o	Omicron	(*Oh*-mih-cron)
δ	Delta	(*Dehl*-tah)	ν	Pi	(*pie*)
ε	Epsilon	(*Ehp*-sih-lohn)	ρ	Rho	(*Roh*)
ζ	Zeta	(*Zay*-tah)	σ	Sigma	(*Sig*-mah)
η	Eta	(*Ay*-tah)	τ	Tau	(rhymes with cow)
θ	Theta	(*Thay*-tah)	υ	Upsilon	(*Up*-si-lohn)
ι	Iota	(*Eye-oh*-tah)	φ	Phi	(*F-eye*)
κ	Kappa	(*Cap*-ah)	χ	Chi	(*K-eye*)
λ	Lambda	(*Lamb*-dah)	ψ	Psi	(*Sigh*)
μ	Mu	(*Myu*)	ω	Omega	(Oh-*may*-gah)

33.3 Abbreviations of constellations

Here is an alphabetical list of the abbreviations of constellation names you may meet. I have suggested pronunciations for some of the more complex names. In the classical method of Latin pronunciation, genitive forms ending in "ae" (Aquilae) are normally pronounced "eye" and those ending in "i" (Cygni) as "ee". In the "English" method, genitive forms ending in "ae" are normally pronounced "ee" and those ending in "i" as "eye."

And	Andromeda	Andromeda
Aps	Apus	Bird of Paradise
Aql	Aquila	Eagle
Aqr	Aquarius	Water Carrier
Ara	Ara	Altar
Ari	Aries	Ram
	(*Air*-ease)	
Aur	Auriga	Charioteer
	Aur-*rye*-ga	
Boo	Bootes	Herdsman
	Boh-*oh*-tes	
Cam	Camelopardalis	Giraffe
	Camel-o-*par*-duh-lis	
Cap	Capricornus	Goat
Car	Carina	Ship's Keel
Cas	Cassiopeia	Queen
	(Cass-io-*pee*-a)	
Cen	Centaurus	Centaur
Cep	Cepheus	King
Cet	Cetus	Whale
Cha	Chamaeleon	Chameleon
CMa	Canis Major	Big Dog
Cnc	Cancer	Crab
CrA	Corona Austrina	Southern Crown
CrB	Corona Borealis	Northern Crown
Cru	Crux	Cross
CVn	Canes Venatici	Hunting Dogs
	(*Cay*-niss Ven-*ah*-tih-see)	
Cyg	Cygnus	Swan
Del	Delphinus	Doplhin
Dra	Draco	Dragon
Equ	Equuleus	Little Horse
	(Eh-*queue*-lee-us)	
Eri	Eridanus	River
	(Air-i-*dan*-us)	
Gem	Gemini	Twins
Her	Hercules	Hercules

Hya	Hydra	Female Sea Monster
	(*Hy*-dra)	
Hyi	Hydrus	Male Sea Monster
Lac	Lacerta	Lizard
	(Lah-*sir*-tah)	
Leo	Leo	Lion
Lep	Lepus	Rabbit
	(*Lee*-pus)	
Lib	Libra	Scales
LMi	Leo Minor	Little Lion
Lup	Lupus	Wolf
Lyn	Lynx	Lynx
Lyr	Lyra	Harp
Mic	Microscopium	Microscope
Mon	Monoceros	Unicorn
	(mo-*naw*-cer-ohs)	
Mus	Musca	Fly
Nor	Norma	Level
Oct	Octans	Octant
Oph	Ophiuchus	Serpent Bearer
	(Off-ee-*you*-cuss)	
Ori	Orion	Hunter
	(Oh-*rye*-on)	
Pav	Pavo	Peacock
Peg	Pegasus	Flying Horse
Per	Perseus	Andromeda's Rescuer
Psc	Pisces	Fish
	(*Pice*-ease)	
Pup	Puppis	Ship's Stern
Pyx	Pyxis	Ship's Compass
Sco	Scorpius	Scorpion
Ser	Serpens	Serpent
Sge	Sagitta	Arrow
	(*Sah*-jit-a)	
Sgr	Sagittarius	Archer
Tau	Taurus	Bull
Tel	Telescopium	Telescope
Tri	Triangulum	Triangle
Tuc	Tucana	Toucan
UMa	Ursa Major	Big Bear
UMi	Ursa Minor	Little Bear
Vir	Virgo	Virgin
Vul	Vulpecula	Fox

Index